ADHESION IN CELLULOSIC AND WOOD-BASED COMPOSITES

NATO CONFERENCE SERIES

I	Ecology
II	Systems Science
III	Human Factors
IV	Marine Sciences
V	Air-Sea Interactions
VI	Materials Science

VI MATERIALS SCIENCE

ADHESION IN CELLULOSIC AND WOOD-BASED COMPOSITES

Edited by

John F. Oliver

Xerox Research Centre of Canada
Mississauga, Ontario, Canada

Published in cooperation with NATO Scientific Affairs Division

PLENUM PRESS · NEW YORK AND LONDON

Library of Congress Cataloging in Publication Data

Main entry under title:

Adhesion in cellulosic and wood-based composites.

   (NATO conference series. VI, Materials science; v. 3)
   "Proceedings of a conference on adhesion in cellulosic and wood-based com-
posites sponsored by the NATO Science Committee (Materials Science Panel) and
held May 12-15, 1980, at Queen's University, Kingston, Ontario."
   Bibliography: p.
   Includes index.
   1. Composite materials—Congresses. 2. Adhesion—Congresses. 3. Cellulose
—Congresses. 4. Wood—Congresses. I. Oliver, John F. II. NATO Science Com-
mittee. Materials Science Panel. III. Series.
TA418.9.C6A266   620.1'292   81-11983
ISBN 978-1-4684-8985-9   ISBN 978-1-4684-8983-5 (eBook)   AACR2
DOI 10.1007/978-1-4684-8983-5

Proceedings of a Conference on Adhesion in Cellulosic and
Wood-Based Composites sponsored by the NATO Science
Committee (Materials Science Panel) and held May 12-15, 1980,
at Queen's University, Kingston, Ontario, Canada

© 1981 Plenum Press, New York
Softcover reprint of the hardcover 1st edition 1981
A Division of Plenum Publishing Corporation
233 Spring Street, New York, N. Y. 10013

CONTENTS

WORKSHOP SESSIONS

INTRODUCTION

Cellulose is a versatile and renewable natural resource which has attracted increasing attention in the last decade, expecially after the energy crisis of 1973. Apart from its extensive use as a solid product, wood is the most important source of cellulose fibres for papermaking and is also widely used as a source of energy. The form and availability of the forest provides a great opportunity for technological improvement and innovation in the future to satisfy the foreseeable increasing demand for wood-based products. For example, North American sawmills and plywood mills presently recover only about 45 to 55% of logged wood while the remainder is disposed as waste, if it is not used in pulp manufacturing. In addition, top and branch wood, and logs from non-commercial species which are presently not recovered from the logging sites could provide an abundant and relatively inexpensive resource for the manufacture of composite products. Other valuable potential sources of cellulosic materials are waste paper and agricultural waste.

A composite is the consolidation of two polymeric materials such that one of the components acts as the adhesive binder while the other forms the substrate matrix. In some cases, the matrix and the adhesive may be the same materials. To maximize the adhesion potential of the composite, the properties of the substrate which can enhance, hinder or complicate the development of optimum adhesion should be thoroughly explored and identified. In cellulosic and wood-based composites, the role of the matrix and the interface and component compatibility are of major importance.

Recent research interest into the use of natural products such as lignin, bark and foliage for adhesives, while achieving some degree of technical success, has also generated controversy over product durability and economy. The research and application of natural product adhesives needs further development and thought, especially in terms of availability, variability and potential bond durability. Moreover, the role of synthetic resins (both thermosetting and thermoplastic) in wood and cellulosic composites should not be overlooked and suggestions to increase their efficiency and diversity for enhancing composite properties are required.

The rheology of composite-product formation using different substrates presents an exciting challenge, and the morphology and mechanical properties of the multitude of wood composite products now in use require additional study. Furthermore the performance requirements of any new composite product needs more thorough examination on a scientific basis in order to minimize the time involved in establishing standards or specifications for product acceptance.

There may be a need for a more liberal classification of the adhesion forces in composite products, emphasizing the adhesive used and the durability required in the end-product. In this way, different categories of composites can be defined. Finally, there should be more emphasis on the economic significance of research results.

To encourage strategic basic research on the problems of formulating cellulosic or wood-based composites, the NATO Science Committee has sponsored this conference with the following objectives:

1.   Identify the scientific factors which:

   a.   Limit the more extensive use of low-cost cellulosic
         materials in composites.
   b.   Restrict the wider use of cellulosic or wood-based
         composites.

2,   Determine the extent of existing knowledge on the
      subject.

3.   Identify the most pertinent problems that restrict the
      development of suitable composites.

4.   Suggest feasible research approaches to solve the
      problems identified.

Successful formulation of composite products depends upon overcoming problems such as the influence of moisture, unfavourable chemical interactions of adhesive and matrix, morphology of components, stability under ambient and varying conditions, economics, etc. Because of the need to focus on and facilitate discussion of these various factors, four groups were organized to deal with the following specific areas:

Surface and Interface;

Morphology and Mechanical Properties;

Durability; and

Fabrication, Applications and Economics.

It is hoped that the outcome of this conference will stimulate the academic and industrial sectors to undertake more research aimed at conserving scarce resources and encouraging greater use of products made from more abundant materials such as wood.

S. Chow
Chairman, Program Committee

PLENARY SESSION I

COMPONENTS

MOLECULAR AND CELL WALL STRUCTURE OF WOOD

Richard E. Mark

Empire State Paper Research Institute
State University of New York
College of Environmental Science and Forestry
Syracuse, N. Y.   13210

1.   THE FIBER COMPONENT OF WOOD - ITS IMPORTANCE TO WOOD PROPERTIES

To the materials scientist, the word "fiber" connotes a slender, thread-like structure with substantial mechanical properties. The term includes a vast array of natural and synthetic materials - derived from metals, polymers, ceramic and vitreous materials, and from diverse animal and plant origins. Wood is a fibrous material in the main.   In the woods produced by conifers, some 90 per cent of the volumetric composition and approximately 95 per cent of the dry mass can be attributed to cellulosic fibers known as tracheids. Lignified cellulosic fibers of various kinds form the bulk of virtually all woods. Accordingly, our attention focuses on the constituent fibers when we consider the material wood from a materials science standpoint.   In wood as well as in other plant-derived materials, the fibers are in reality the exoskeletons of once-living, elongate cells that originated from a meristem.   The cells each have a lumen, or hollow center.

Figs. 1 through 3 are scanning electron micrographs of a small block of pine wood.   The cell elements are identified on the photographs.   By observing these illustrations, one can readily appreciate how wood gets its grain direction from the oriented packing of the fibers, how density is related to the ratio of cell wall to lumen volume, and why wood conducts heat, electricity, liquids and gases anisotropically.

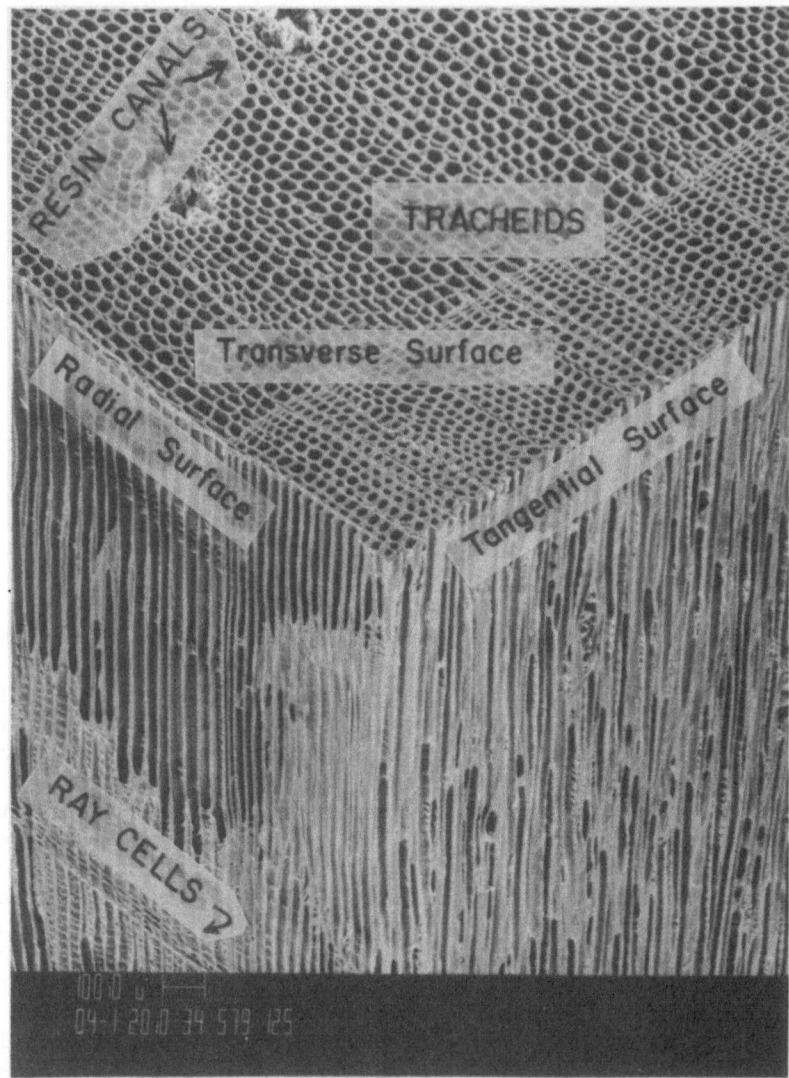

Fig. 1   A 3-dimensional scanning electron micrograph of a conifer
wood, eastern white pine. Parts of 2 annual rings are included. Most of the cellular elements in this wood are hollow, tubular fibers known botanically as tracheids. The
grain of wood (vertical in this picture) arises from cellular alignment. On the surface cut transverse to the grain
one can see 2 areas of apparent disruption. These areas
are actually resin canals with associated small resin-
producing cells around them. Resin canals are sometimes
visible to the unaided eye. The bar scale at the lower
left represents 100 μm. Micrograph courtesy Dr. W. A.
Côté, Jr.

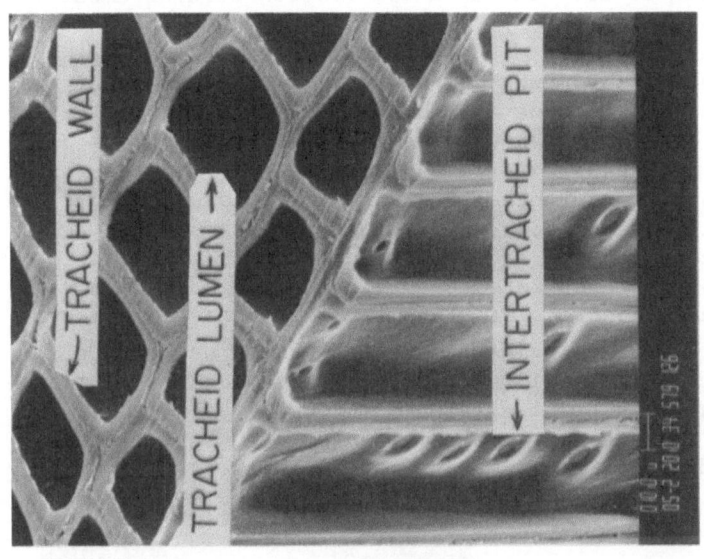

Fig. 3

High magnification scanning electron micrograph of eastern white pine wood. In this
view it is evident that the wall thicknesses
of adjacent tracheids tend to be comparable.
Bar scale represents 10 µm. Micrograph
courtesy Dr. W. A. Côté, Jr.

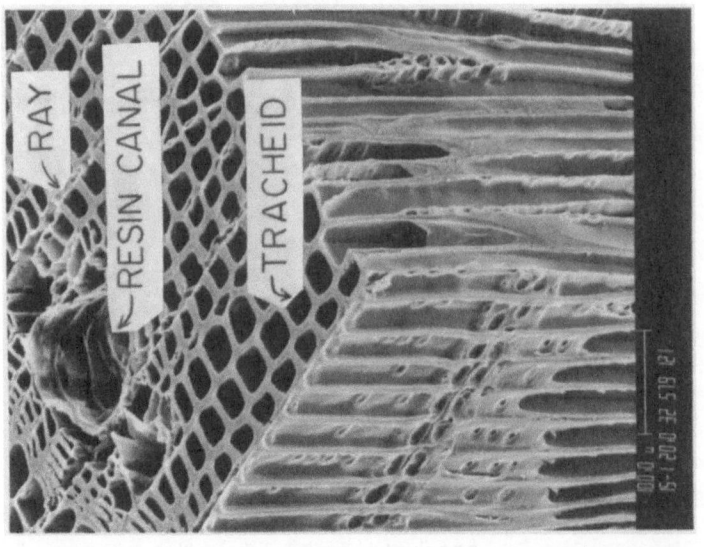

Fig. 2

Three-dimensional scanning electron micrograph
of eastern white pine wood at higher magnification than Fig. 1. Additional details of the
tracheids, resin canals and other cellular
elements are evident. Bar scale represents
100 µm. Micrograph courtesy Dr. W. A. Côté, Jr.

2.   THE MOLECULAR COMPONENTS OF WOOD FIBERS

One may think of the cell wall of a wood fiber as a composite material, for indeed it is. It has a filamentous reinforcement material (principally cellulose), a matrix of polymers that holds these filaments in place, and a few fillers in the matrix that modify matrix properties somewhat. Structurally, the simplest way to envision how the filamentous reinforcement relates to the cell wall of the fiber is to imagine that you have a skein of yarn looped in a loose coil hanging from your fingertips. Now take the bottom of the loop with your other hand and give the bundle of yarn a twist as you pull on it. You are left with something that re-sembles a rope except for the end loops that curve between your fingertips (Fig. 4). If you were to twist this skein a few more times and then encase the resultant structure in some rigid matrix material, you would have a rudimentary model of the way a wood fiber is constructed. Your filaments of yarn are analogs of micro-fibrils, the actual structural filaments in wood fibers (refer to Fig. 5). We will consider first the molecular structure of the microfibrillar and matrix constituents, then examine the way that

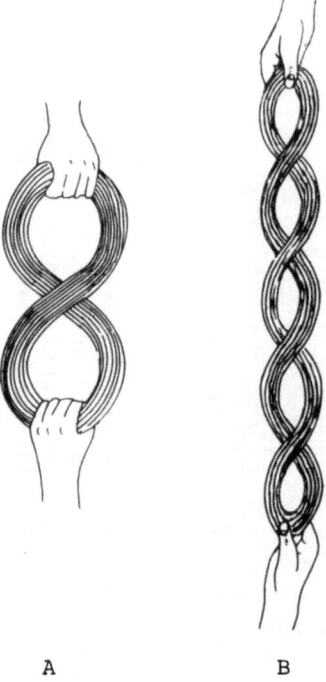

A   B

Fig. 4.   Twisted skein

Fig. 5.   Aggregations of micro-fibrils at the end of a crushed tracheid from which the matrix has been removed. From Bucher[8]

they are assembled together in greater detail. One must distinguish between the terms <u>microfibril</u> and <u>fiber</u>. The microfibril represents just one strand of the "yarn" that makes up the structural framework (the twisted skein) of the fiber cell wall.

2.A.   Physical Nature of the Polymeric and Other Constituents

The most important of the chemical constituents of wood fiber cell walls is cellulose, a long-chain, unbranched condensation polymer of β-D glucose units. The structure of a glucose unit is illustrated in Fig. 6 (A and B). It is a ring structure formed by the bonds between five carbon atoms and one oxygen atom. In Fig. 6A, a planar schematic drawing of the glucose monomer illustrates the α and β isomers of D-glucose. Of the two possible D-glucose structures, one has the hydroxyl (OH) groups on the top two carbon atoms on the same side of the ring (the α form) and the other (β) on opposite sides. The D (for dextro-) notation comes from the convention of also describing sugar molecule isomers by the position of the OH group on the next-to-last carbon atom in the planar projection. The reference OH group in Fig. 6A is to the right (dexter) side of the carbon chain; hence the monomer is D-glucose.

Under certain conditions, solutions of D-glucose will undergo changes from α or β form to some equilibrium mixture of the two via a transient intermediary form in which the ring is broken open, as also shown in Fig. 6A. But the stable form of glucose is the ring with side groups as shown. A ring formed from 5 carbon and one oxygen atoms is said to have a "pyranose" form. Fig. 6B provides a schematic of the α and β ring structures in which the rings are illustrated as planar (they are actually bent). The carbon atoms in Fig. 6B are numbered according to convention. The α and β forms form polymers that are distinctly different in many chemical and physical properties.

In a native cellulosic fiber, the number of cellulose molecules that pass through a given cross section is on the order of one to two billion. The cellulose that is present in wood differs substantially from the "cellulose" that is found in such industrial products as rayon tire cord or cellophane. The differences are in some cases chemical but more usually physical in nature. For this reason, the cellulose found in wood (and other plant tissues) is designated "native" cellulose to distinguish it from the modified industrial celluloses. The supermolecular arrangement of native cellulose (as it is normally found) is referred to as cellulose I.

A very significant physical aspect of cellulose I is that it is extensively aggregated into monoclinic crystalline arrays. These crystalline arrays ("crystallites") are the principal, if not

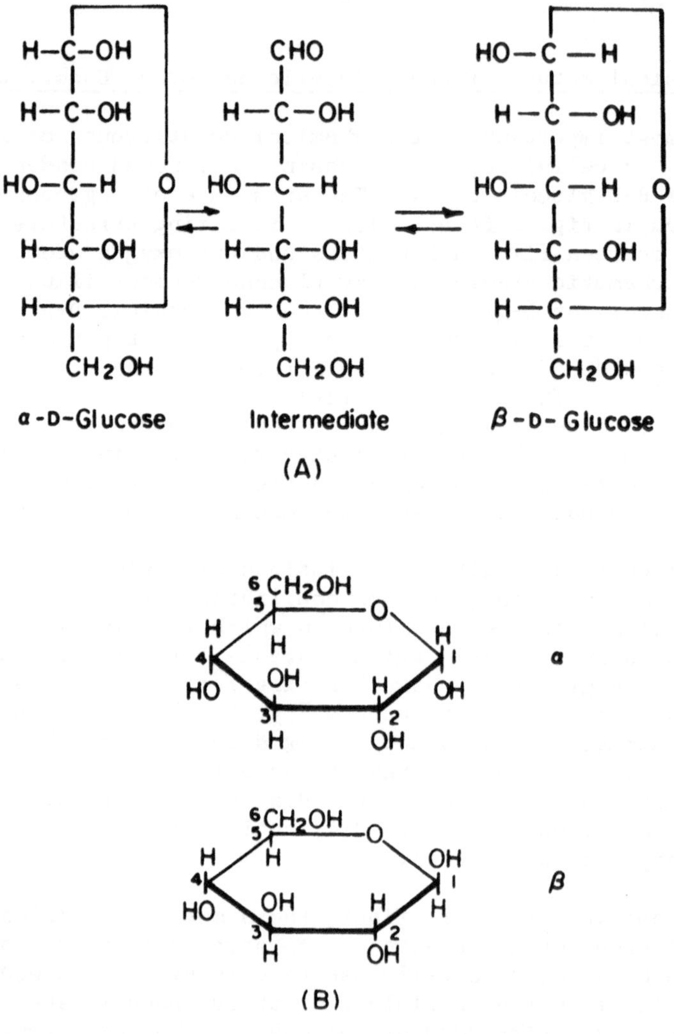

Fig. 6. Structure of glucose. (A) Planar schematic for D-glucose,
         showing the two isomeric forms (left and right) and an
         open-ring intermediary form (center). (B) Planar sche-
         matic of ring structures for α and β D-glucose, showing
         conventional numbering of carbon atoms.

exclusive, constituent of microfibrils, which are the observable
(by microscopy) filamentary structures that form the structural
framework of the wood cell wall.

Until quite recently, it was widely believed that the chain
molecules in cellulose I were antiparallel, i.e., that chains
packed together in a crystallite of native cellulose were arranged
in an alternating pattern as to chain directionality. Conclusive
proof that all chains within the microfibril are oriented in the
same direction has been developed by computer-assisted crystallographic and conformational analysis, and the development of methods
to determine the minimum energy packing arrangement of the molecules.

The typical length of a chain molecule of native cellulose$_o$is
unknown but probably well exceeds the value of 10^4 nm (100,000 Å)
that may be determined experimentally for the substance. The
reason for this statement is that the physical removal of one chain
from its associates in the cell wall (by whatever means) in order
to measure it, almost certainly results in scission (breakage) of
primary valence bonds linking the glucose units together; thus, we
actually measure chain fragments. As chemical techniques for removing cellulose from fibers have become progressively less harsh,
measured values for the degree of polymerization (DP) have risen
from numbers in the hundreds to numbers sometimes exceeding 30,000.

While nearly one half of the dry mass of wood is cellulose,
there is a family of related polymers present whose structures are
based on various combinations of sugar residues of xylose, glucose,
mannose, galactose and arabinose. These other polysaccharides are
collectively known as hemicelluloses. Typically the hemicelluloses
are branched polymers having DP values in the low hundreds. They
vary from relatively unbranched, alkali-resistant species such as
glucomannan to the relatively non-linear, soluble types such as the
arabinogalactans, which can form modified sugar polymers containing
pectic and uronic acid residues. The nature and proportions of the
hemicelluloses found in different woods varies, although there are
broadly consistent patterns. For example, the predominant hemicellulose in hardwoods* is a glucuronoxylan whose structure is
remarkably consistent from wood to wood. In softwoods* the main
hemicelluloses are the galactoglucomannans, a family of closely
related polysaccharides that differ principally in the proportions

---

*The names "hardwood" and "softwood" refer to the botanical origin
of the particular wood and not to the hardness of a particular
wood on either absolute or relative scale. It is true that the
wood of a conifer (softwood) is usually lighter and softer than a
typical hardwood. However, there is great variability in both
categories, expecially the latter. Thus some of the world's
softest woods are hardwoods.

of their component sugar residues. There are two types normally
found - a relatively soluble polymer containing mannose, glucose,
and galactose units in the ratio 3:1:1 and a more more resistant
type found in close association with cellulose, in which the ratio
approximates 30:10:1. Other galactoglucommans with somewhat different polymer compositions may also be present. Hemicelluloses
may exhibit some degree of orientation and crystallinity, particularly when they are in close association with cellulose.

An amorphous polymeric material called lignin comprises the
third major molecular component found in all woods. As with the
hemicellulose group, the actual composition of "lignin" varies between woods even though there are fairly consistent general differences between the lignins in hardwoods as compared with softwoods.

Lignins are polymerized enzymatically in plant cell walls from
three primary precursor monomers: with reference to Fig. 7, these
precursors are (1) trans-coniferyl, (2) trans-sinapyl, and (3) trans-
p-coumaryl alcohols. These alcohols form the most common polymer
structures found in various lignins. The polymerization is dehydro-
genative in nature. A typical conifer lignin might be composed of
units such as the 16-monomer unit shown in Fig. 8. The majority
of these units are of the guaiacyl propane type, which derives from
the trans-coniferyl monomer. Unit No. 13 exhibits two methoxy
(OCH_3) side groups; it is a syringyl propane unit, derived from
trans-sinapyl alcohol. The derivative of the trans-p-coumaryl unit,
the third precursor in Fig. 7, is coumaryl propane (unit No. 2 in
Fig. 8).

In spruce, which has a rather typical conifer lignin structure,
the lignin originates from coniferyl, sinapyl, and coumaryl alcohols in a ratio of approximately 80:6:14. The lignin of beech,
which is fairly typical for hardwood, has ratios of 49:46:5. Thus,
the softwood lignins are predominantly of the guaiacyl type, while
guaiacyl and syringyl groups are about equally represented in the
hardwoods.

Just as removal of native cellulose chain molecules in an
intact condition from a fiber cell wall is physically difficult if
not impossible, it has not yet been possible to remove lignin from
wood without some alteration of the chemical structure during
contact with the solvent. The molecular weights of soluble lignin
derivatives cover a range of over 3 orders of magnitude, typically
($<10^3$ to $>10^6$). Goring[16] notes that, "Apparently, lignin can be
dissolved as an entity small enough to be a pure chemical compound
or as a particle large enough to show the behaviour of a high polymer of a colloid....such wide polydispersity leads to a certain
degree of indeterminacy in ascribing a single value of the molecular weight to a given sample of lignin. A number average

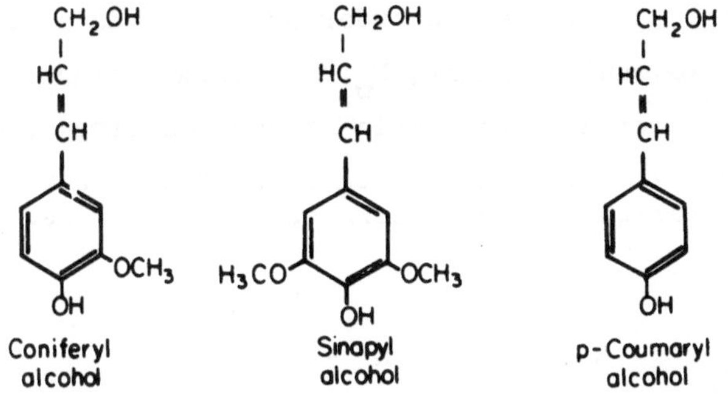

Fig. 7. Precursor monomers of lignin (24)

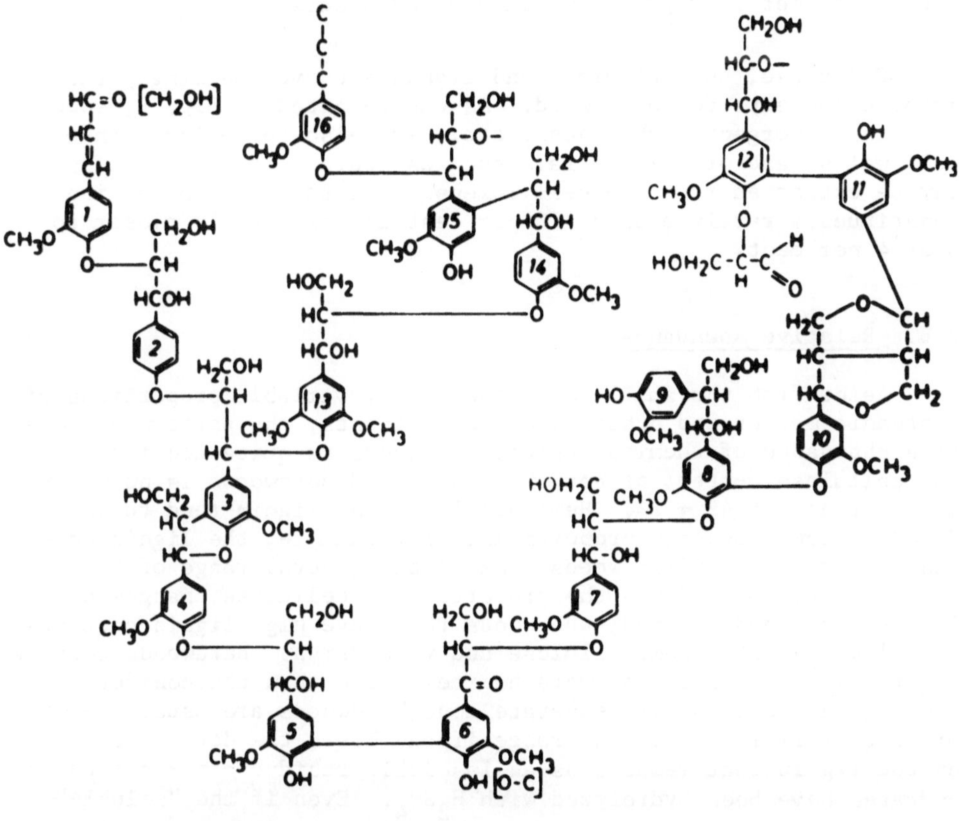

Fig. 8. Proposed model for a typical softwood lignin (1).

molecular weight, M_n, may be an order of magnitude smaller than the weight average molecular weight, M_w." Determination of both M_w and M_n and computation of the ratio of the two gives a measure of the polydispersity of the lignin fractions that are isolated. Average values for these two numbers can be determined from

$$M = \frac{\sum_i w_i}{\sum_i w_i / M_i} \qquad\qquad M_w = \frac{\sum_i w_i M_i}{\sum_i w_i}$$

wherein the w_i values are the weights and M_i values are the molecular weights of the individual fractions obtained by some technique such as dialysis or precipitation. Experimental results show that M_w/M_n ratios of between 2 and 7 are found for the lignins removed from various woods. For a monodisperse system, $M_w/M_n = 1$. If a linear polymer is degraded by random chain scission, $M_w/M_n \rightarrow 2$.

Cellulose, hemicellulose and lignin are overwhelmingly the major components found in wood. As can be noted in Fig. 9, some proteins, inorganic substances, etc., are also to be found in wood cell walls, and other materials such as starch, fats and resins may be extracted from the cell lumens. The total of these minor constituents rarely exceeds 10 per cent and is usually closer to 3 or 4 per cent.

2.B. Relative Abundance

Aside from the small and often highly variable proportions of inorganic matter and other minor constituents, the pattern of relative abundance of chemical species in woods is quite consistent. The cellulose content of both hardwoods and softwoods is normally in the range of 42 ± 2%. Hemicelluloses and lignin tend to be found in complementary proportions. For example, the lignin content of various conifer woods lies in the general range of between 24 and 33%. Since the proportion of cellulose changes but little from wood to wood, the woods that have high lignin contents have low (ca. 25%) hemicellulose and vice versa. Hardwoods contain generally less lignin and more hemicellulose than the conifers. The lignin contents for temperate zone hardwoods are usually determined to fall in the 16-24% range if one bases the determination on the lignin that remains as an insoluble residue after the carbohydrates have been hydrolyzed with H_2SO_4. Even if the "soluble" lignin is counted, the above hardwood lignin percentage only

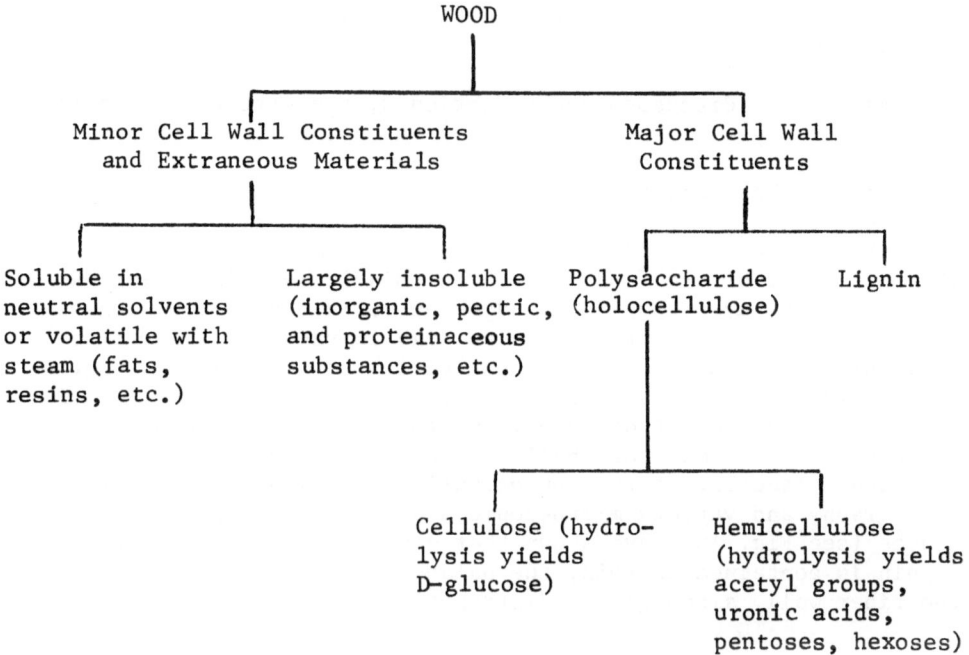

Fig. 9. Constituents of wood (adapted from Ref. 7).

rises to the range of 19-28.

While it has been noted above that the minor constituents usually constitute only 3 or 4 per cent of the mass of wood, some tropical hardwoods are found that contain 5 per cent or more of inorganic materials such as SiO_2 and MgO, which quickly dull cutting tools. Also, certain woods (both conifer and angiosperm)* contain substantial amounts of extractives such as organic crystals that have toxic or repellent effects on organisms that attack wood (e.g., insects, fungi and marine borers). Thus, the minor constituents may have significant effects on industrial and other applications of certain woods, even though they appear in relatively small quantities.

Abundance of various molecular species also varies within a species of tree because of soil fertility and other environmental factors, and within the wood of one individual tree. As an example of the latter, the lignin content of the fibers in one annual ring of a temperate zone tree tends to be higher in the fibers formed

---

*conifer $\simeq$ softwood; angiosperm $\simeq$ hardwood

at the start of the growing season ("springwood") than in those
formed later ("summerwood").

     As will be discussed under Section 3, the structure of a fiber
consists of layers. The abundance of various polymer species and
minor constituents also varies from layer to layer. While polysaccharides that contain glucose, mannose and xylose residues tend
to be distributed throughout the cell wall, the galactan, arabinan
and other pectopolyuronide polymers are typically concentrated in
the outer part of the fiber and in the middle lamella, which is
the cementing layer that bonds adjacent cells together. Although
the concentration of lignin is also high in the middle lamella,
that structure does not usually exceed 12% of the volume of normal
wood. Since the middle lamella is less dense than the cell wall
and the lignin content of wood is on the order of 20 to 30%, it
is apparent that less than half of the total lignin is located
the middle lamella. Actual quantitative determinations by micro-
radiography and various microscopic and microspectrographic techniques (see Fig. 10) have shown that some 15 to 25% of the total
lignin is contained therein, the balance being found throughout
the fiber wall, although the distribution varies among the layers.

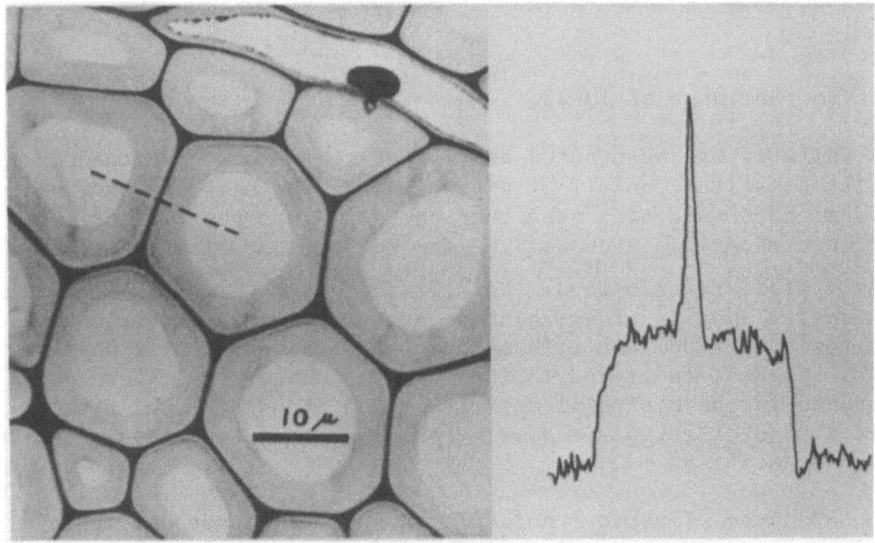

Fig. 10. Shown are wood fibers in cross section and a curve of the
lignin distribution across the double cell wall. The distribution
curve is obtained by a quantitative microspectrographic technique
applied along the dashed line superposed on the photograph. Photo
and curve courtesy of Dr. D. A. I. Goring (from Ref. 10).

2.C.   Aggregation, Juxtaposition and Conformation

In Figs. 4 and 5 a rudimentary concept of the microfibrillar constitution of wood fibers was presented. The microfibril is composed principally, and in some plant cells exclusively, of cellulose I. The orientation of the cellulose chains has been shown to be the same as the microfibril. Views of microfibrils in a cellulose-containing seaweed are given in Fig. 11. The visible (by electron microscopy) cross-sectional dimension of a microfibril is typically on the order of 10 nm × some multiple of 3.5 or 4 nm, and some scientific workers consider this size to represent the basic supermolecular structural unit. However, a variety of evidence has been presented to support the concept that the microfibril in turn is composed of more slender sub-units (ca. 3.5 nm × 3.5 nm). Such "elementary fibril" sub-units have now been derived by a number of chemical, enzymatic and physical techniques (see Fig. 12) although in most cases the results are open to much interpretation and are still controversial.

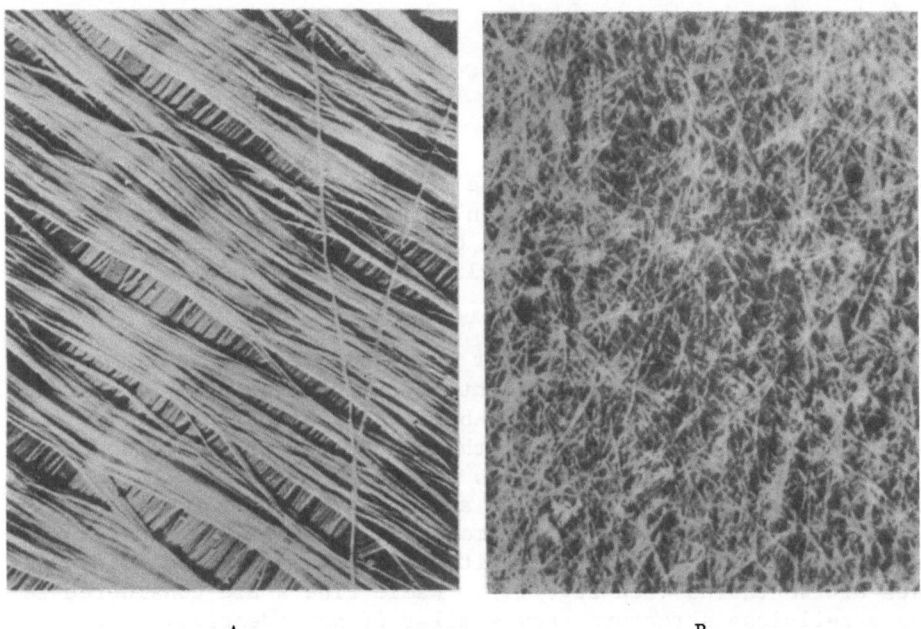

A   B

Fig. 11. Microfibrils of the cell wall of the marine alga Valonia. Most of the wall possesses a microfibrillar texture similar to that appearing in A, but the pattern exhibited in B is typical of the outermost part of the cell wall. Wood fibers possess homologous microfibrillar textures. Photomicrographs courtesy of Dr. K. Mühlethaler (from Ref. 11).

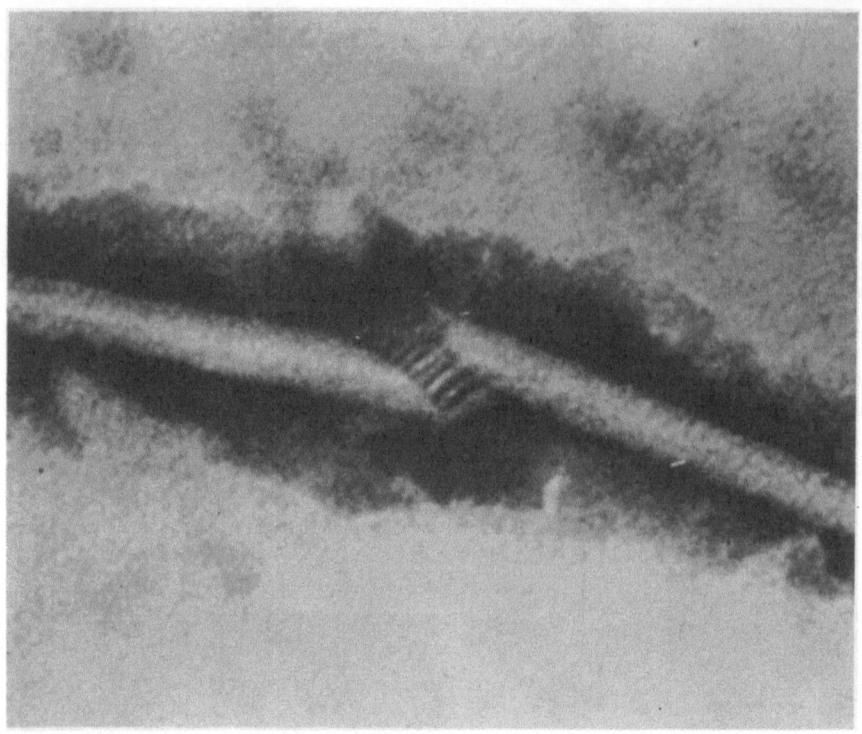

Fig. 12 Microfibril of cellulose subjected to ultrasonication.
        The structure has broken into sub-units approximately
        3.5 nm wide in the area of the kink. Photomicrograph
        courtesy of Dr. R. Muggli (from Ref. 12).

        Within the microfibril and/or elementary fibril, chains of
cellulose I are packed together into a crystalline array of parallel
chains. Although it is generally believed that cellulose I is the
same (chemically and physically) throughout the plant kingdom, the
evidence to support that belief is somewhat ambiguous. The crystal
symmetry of cellulose I in ramie, an important fiber for paper,
cloth, and cordage, is certainly monoclinic ($a \neq b \neq c$, $\alpha=\beta=90°$, $\gamma \neq 90°$).
The space group is $P2_1$ and the unit cell contains disaccharide seg-
ments of two chains (see Fig. 13). Ramie fibers are considered to
have a cellulose structure very similar if not identical to that
found in wood.

        Sometimes when cellulose I is studied in other plant mater-
ials (e.g., by X-ray methods), certain reflections indicate that
there must be 8 chains in the unit cell. Also, it appears that the
monomeric (glucose) residues in some cases are staggered in a
manner that results in a triclinic ($a \neq b \neq c$, $\alpha \neq \beta \neq \gamma \neq 90°$) crystal symmetry. But in such cases there is other evidence to the contrary,

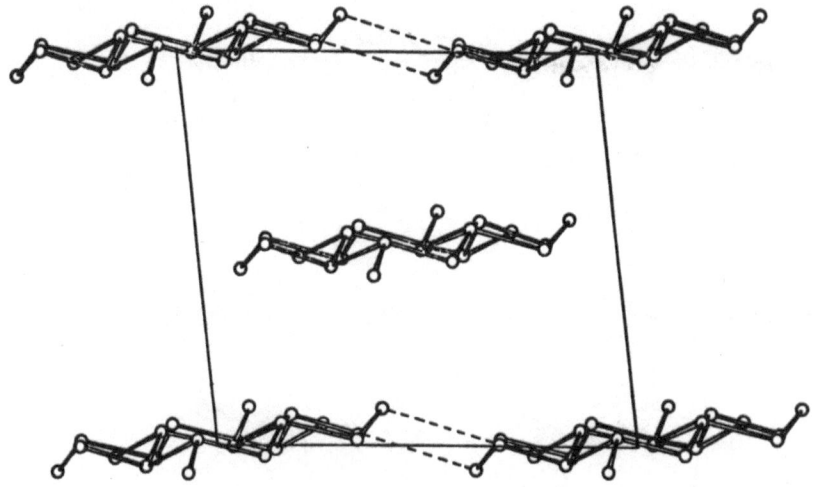

Fig. 13.   Projection of a 2-chain cellulose I unit cell viewed
           perpendicular to the ab plane (along the fiber axis),
           The suggested hydrogen bonding pattern for ramie cellu-
           lose is shown (dashed lines). This picture, reprinted
           with permission from Woodcock[28], was rephotographed so
           that the lattice parameters follow the same convention
           as shown in Fig. 14. It should also be noted that the
           designations of the a and b axes used by Woodcock differ
           from those shown in Fig. 14. That is, the b axis is
           horizontal in this figure.

and it is not generally possible to accept one concept or another
conclusively. In Fig. 14, an array of cellulose chains is shown.
Three different possible unit cells are illustrated by this drawing.
Possibly they are all correct for cellulose I for different plant
origins. For the purposes of this discussion, we accept the description of Woodcock[28], who designates the unit cell as monoclinic
with dimensions a = 0.778 nm, b = 0.820 nm, c = 1.034 nm (chain
axis) and γ = 96.5° (see Fig. 13).

     X-ray line broadening measurements and other techniques indi-
cate that the microfibrils are essentially single crystals, at
least in certain algal cell walls that contain cellulose I in a
very pure form. However, the situation is not always so well defined. In conifer wood cell walls there is a small amount of glucomannan hemicellulose that seems to be very intimately associated
with the exclusively microfibrillar, highly crystalline native

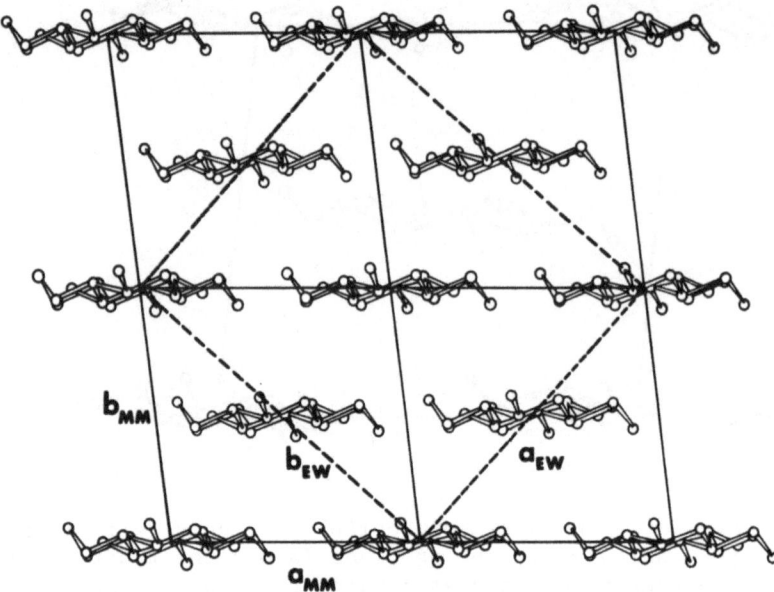

Fig. 14.   Projection of the crystal structure of cellulose I.
           Three structural possibilities are shown, a 2-chain (MM)
           unit cell, an 8-chain unit cell that is 4 times the size
           of the MM cell, and a 4-chain (EW) unit cell (from Ref. 5).

cellulose after the other cell wall components have been extracted
(i.e., the lignin, most of the hemicelluloses, and the minor constituents). This tenacious hemicellulose component cannot be removed without degradation of the microfibrils and therefore is
integral with them.

    In Fig. 15 a schematic view of a microfibril in its environ-
ment in a living wood cell is shown. In a space 12 nm × 43 nm
(these dimensions are highly variable) are shown 6 microfibrils
each 10 nm × 3.5 nm. The microfibrils contain the crystalline
polymer of β-D glucose residues (cellulose), with or without some
occasional substitutions of mannose for glucose residues in random
or periodic form to introduce crystal order defects. The normal
linkage of these residues is for a bridge oxygen atom to join the
carbon atom at position 1 on one residue to the carbon atom at
position 4 on the next (see Fig. 6B for numbering of the carbon
atoms). This normal 1.4 linkage might also be occasionally altered
along the length of the chain by 1,3 linkages, which would thereby
create some local disorder.

    Also in Fig. 15 one can observe extractable short hemicellulose

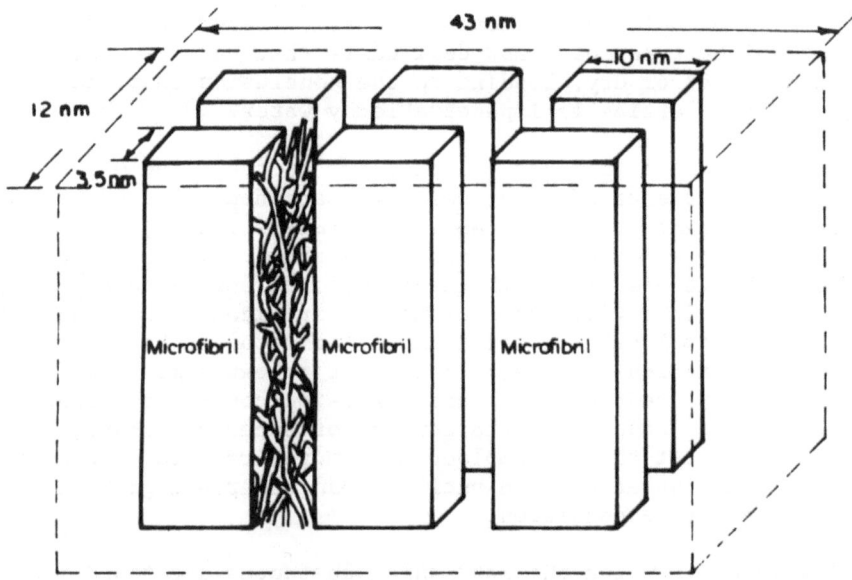

Fig. 15. Diagrammatic representation of a small element of wood
cell wall. Six microfibrils are present. Between two
of the microfibrils, bent and branched rods representing
hemicellulose chains are shown. Hemicelluloses, as well
as the other substances present in wood, occupy the
spaces between microfibrils throughout the cell wall.

chains between the microfibrils. The hemicelluloses also exhibit
some degree of orientation but have considerably less crystallinity
than the cellulose. Other substances (lignin, proteins, inorganic
matter, air and especially water) occupy the remainder of the
12 nm × 43 nm space. The architecture of the woody plant cell
wall then consists of many of these spaces. Dimensions may vary,
and there may be considerably less alignment of the hemicellulose
than shown in Fig. 15, but this general scheme appears to hold
for the entire spectrum of lignified cellulosic plants.

It has been shown that the removal of hemicelluloses, lignin
and minor constituents has no effect on the X-ray diagram of a
cellulosic cell wall so long as the cellulose is not degraded
during extraction. A minor exception to this is when occasionally
some increase in crystallite size (sharpening of the arcs) occurs,
suggesting perhaps that the alignment of chains on the microfibril
surfaces has become more regular. When water is removed from the
cell wall, it is observed that the wet walls give identically the
same X-ray diagram as do dry walls except for the presence of an

overlying water halo in the wet wall case. The crystallite sizes, at least in the types of plant cell walls that have been studied[9,23] are the same wet or dry, leading to the conclusion that the cellulose I crystal lattice is impenetrable by water.

It can be seen from Fig. 15 that two of the three major solid chemical constituents of wood, cellulose and hemicellulose, tend to be linear and tend to be aligned (the former strictly, the latter less so) with the microfibril axis. The third major solid constituent, lignin, is highly three-dimensional and as noted before, may be present in a wide range of molecular sizes. The shape of a soluble lignin molecule in solution conforms to the behavior expected of a spherical microgel in respect to experimentally determined viscosity, sedimentation and diffusion constants. Electron microscopy lends support to the concept of spherical shape, but other data suggest the macromolecule is more irregular and asymmetric; we will never be sure until we can observe lignin in its natural solid state condition.

The forms in which the inorganic substances are found in wood vary. Metallic salts may have varying degrees of crystallinity, but quite often materials such as SiO_2 contain water of hydration and are found as amorphous "skeletons" on and throughout the cell walls. Proteins and the extractable or relatively soluble materials tend to be localized, the proteins being mainly in the middle lamella and the extreme outer part of the wall, and the other materials on the inner surface and in the lumen of the cell. There are exceptions to this pattern, however.

The overall picture of a unit volume wood cell wall at the supramolecular level is

(a) A crystalline, filamentous, solid phase material (the microfibril) composed nearly exclusively of cellulose I and which is impenetrable to water, accounts for approximately 42 per cent of the dry solid mass, but a somewhat lower proportion of the total volume under normal ambient conditions because of the presence of air and water.

(b) Two interpenetrating solid phase systems, one composed of an extensively branched, three-dimensional, amorphous polymer (lignin), and one composed of a complex of relatively linear (but nevertheless partially branched), partially paracrystalline polymers of a variety of molecular sizes and solubilities (hemicelluloses), account for most of the remaining solid fraction and create a matrix of polymer materials in which the filamentous microfibrils are embedded.

(c) A third very delicate and tenuous interpenetrating system of solid inorganic matter exists, in addition to the lignin and

and hemicelluloses, which remains as a recognizable structure when wood is carefully ashed. It comprises but a small proportion, on the order of 1 percent, of the dry mass of wood.

(d) A fourth and fifth interpenetrating system can be ascribed to the presence of water and air within the cell wall structure. Both water and air are always present, but their proportions vary with the dryness of the ambient environment. If little water is present, as in the case of wood in a low-humidity environment, the water will only exist in the form of (a) water of constitution and (b) adsorbed water, wherein the H_2O molecules are hydrogen-bound to the surfaces of the carbohydrate and protein molecules present. This binding may be monomolecular or polymolecular. If the wood is wet, or subject to extreme conditions of humidity, there may additionally be free water present, which penetrates and swells the spaces between microfibrils. Air, or in general gas, occupies space within wood in those spaces not occupied by solid or liquid substances.

(e) Substances such as fats, starches, resins, gums, organic crystals and other extractives, and proteins are usually not regular components of the spaces or unit volumes in wood cell walls but exist in specific locations as inclusions or depositions; except for the proteins and some extractives, they are typically found only in cell lumens and intercellular spaces and are thus not considered as intrinsic molecular components of wood fibers and other cells in wood.

2.D.   Structural Functions of the Constituents

From the foregoing, one can appreciate that any unit volume of wood cell wall on the size order shown in Fig. 15 can be described as a filamentary composite. Since the fiber (or other cell) wall is built up of layers of such filament-reinforced matter, it is also accurate to describe it as a laminated composite.

Based on experimental evidence (direct and indirect) and deduction from experience with analogous industrial materials and structures, we can assign structural functions to each of the components found in a typical unit volume of fiber wall. As with industrial multi-phase materials, certain of the components in wood cell walls act as a structural framework (reinforcement) and others act as the bulking and stiffening matrix in which the reinforcement is embedded. In fibers and other plant cell walls, the microfibril provides the structural reinforcement. And since the microfibril is principally if not exclusively cellulose I, the mechanical properties of that crystalline substance are used identically for the properties of the microfibril. We do recognize, as has been mentioned earlier, that in some woods at least part of the glucomannan

hemicellulose is intimately associated with native cellulose in
the microfibril and must therefore also be considered part of the
reinforcing phase, or framework, of the wall. However, up until
now, computations for the elastic constants and other physical and
mechanical properties of the solid state, microfibrillar entity
have been restricted to dealing with it as if it were composed
exclusively of crystalline cellulose I, without any defects. Also,
these computations have been done on the basis of an antiparallel
chain packing configuration because the discovery that all chains
are parallel is quite recent.

The matrix phase in a cellulosic fiber cell wall consists of
the relatively unoriented or amorphous, short-chained or branched,
polymers of various molecular species (i.e., lignin and hemicellulose and any other polymers such as pectopolyuronides that might be
present in a given layer) plus all of the tiny voids and the gases
and adsorbed water associated with them – in other words, everything
else that surrounds the microfibrils in their local environment.
As properties of the non-cellulose molecular species and/or water
content change, the matrix constants will change, whereas undegraded
cellulose I is considered invariant, at least within constant temperature conditions.

As mentioned earlier, cellulose I has a monoclinic crystal
structure characterized by a repeat distance of 1.034 nm (two anhy-
droglucose units) in the chain direction ($\underline{c}$ axis). The spatial
arrangement of the chains is shown in Fig. 16. An important feature
of the crystalline structure from the standpoint of mechanical,
electrical, etc., properties is the pattern of hydrogen bonding,
since there are no interchain primary bonds. Lateral hydrogen
bonds provide a mechanism for stabilizing the crystal against relative displacement of the chains in response to imposed physical
forces. One may think of the hydrogen bond network as a sort of
cross-bracing. These bonds greatly inhibit any tendency for trans-
lation or rotation of the chains within the crystalline microfibril.

Among all properties, the set of elastic constants applicable
to a given structural solid material are generally the most important. These elastic constants may be measured physically in laboratory tests on materials that are available in the form of large
specimens. For microcrystalline materials such as cellulose I,
these elastic constants may be derived on the basis of known crystal
structure and the energetics of bond stretching and bond angle
change within that structure.

The sum of the energies associated with deforming each bond
length and bond angle within the cellulose I unit cell are equated
to the overall elastic strain energy of the unit cell itself. In
its simplest form, we can make an assumption that a unit cell of
the cellulose I crystal undergoes a change in length in the direction

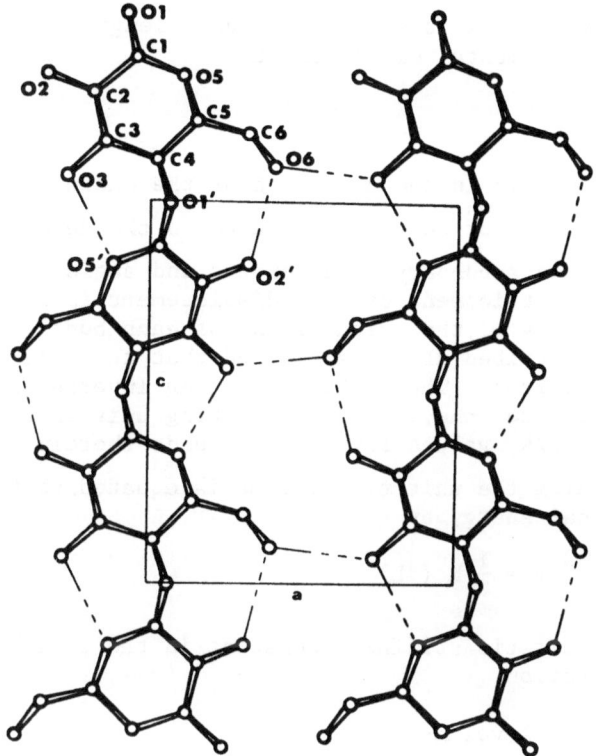

Fig. 16. Projection of the (020) plane of cellulose I showing the
          hydrogen bond network. Each glucose residue forms two
          intramolecular bonds (03 – H···05' and 02' – H···06) and
          one intermolecular bond (06 – H···03). From Gardner and
          Blackwell[13].

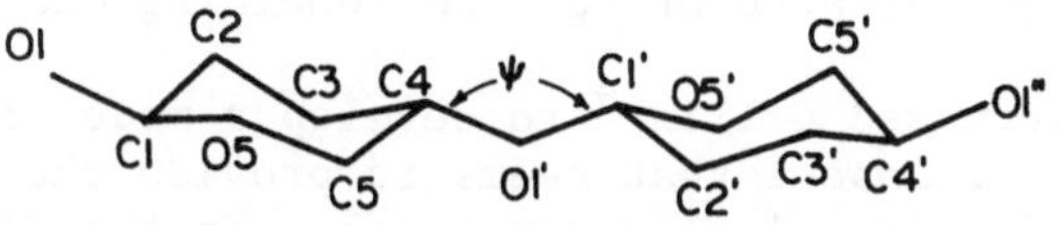

Fig. 17. The basic skeleton of the repeating unit (2 glucose resi-
          dues) in cellulose. Ring and bridge carbon and oxygen
          atoms are shown, but side groups are omitted for simplicity.

of the chain axis so that the repeating unit length L = 1.034 nm
experiences an incremental displacement

$$dL = Le_3$$

where e_3 is the strain in the direction of the chain axis. The
axial deformation Le_3 of course is related to the deformations of
all the primary and secondary bond lengths and angles within the
unit cell. An apportionment of this displacement is made (vecto-
rially) on the basis of the orientations of each bond length and
angle. For a prescribed dL it is assumed that the bond deformation
energies are minimized. The method yields an inverse gross spring
force constant 1/K for the 1.034 nm repeating unit as the sum of
terms containing $1/K_i$ values for all the bonds therein. Then, the
energy in straining the unit cell volume is equated to the overall
bond deformational energy by

$$W = \frac{1}{2} C_{33} e_3^2 \Delta V = \frac{1}{2} K (dL)^2$$

where C_{33} = the elastic stiffness constant in the chain axis
                direction

       ΔV = unit cell volume

Solution of the equation yields

$$C_{33} = \frac{KL^2}{\Delta V}$$

It is possible to derive a full set of elastic constants for
crystalline cellulose by this method[20]. Also, by examining $1/K_i$
terms solved with appropriate numerical substitutions, one can
ascertain where the greatest compliance is occurring in the chain
repeat unit. It has been found that about one-half of the total
compliance arises from the changes that occur in the valence angle ψ
at the bridge oxygen atom. All other bond distentions and bond
angle changes together account for the remaining one half.

This discovery led Gillis[14] to develop a model for the internal
deformation of cellulose I that seems to provide the best estimate
we have to date for its elastic constants. In the Gillis model,
lateral restraint is imposed on the deformation of the valence angle
of the bridge oxygen (the angle formed by C4-O1'-C1' in Fig. 17)
by assuming that each bridge oxygen is hydrogen-bonded to a neighbor
chain. When the bridge oxygen angle changes, the hydrogen bond

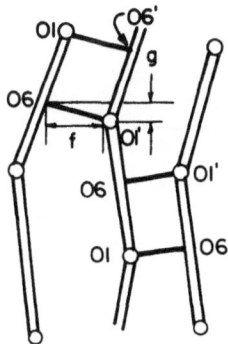

Fig. 18.   Schematic representation of inextensible glucose units
           linked within a cellulose chain by bridge oxygen atoms
           having deformable valence angles and linked by hydrogen
           bonds in the interchain directions[14].

associated with that oxygen must undergo a change in length, and
this has a major effect on the oxygen angle change, and thus the
compliance. The reason for this lies in the fact that stretching
force constants of secondary (hydrogen) bonds are of the same order
of magnitude as the bending force constants of primary bonds. To
account for this restraint, Gillis devised the segmented chain model
shown in Fig. 18.

     Each bar in Fig. 18 represents the length of a glucose residue.
All primary bonds except the bridge oxygen angle are assumed to be
rigid. Under axial tension, the chains tend to straighten out because the distances between the bridge oxygens (e.g., the 01 to 01'
length in the center chain in the illustration) stay constant. The
calculations that have resulted from the assumptions on which this
model is based yield values for stiffness constants C_{ij} and technical moduli (engineering elastic constants) which are summarized
in Table 1. Solution of the applicable equations requires reasonable assumptions for the force constants of hydrygen bonds. The
values in Table 1 were obtained using K_s = 30 N/M and K_b = 8.5 ×
10^{-20} N·m/rad. These are the stretching and bending force constants, respectively.

     In the latest refinements offered by Woodcock[28], Sarko and
Muggli[25], and Gardner and Blackwell[13], the acceptable hydrogen

Table I*

(adapted from Ref. 20)

Stiffness Constants and Technical Moduli for Cellulose I
Matrix of Stiffness Constants (GPa)

$$[c_{ij}] = \begin{bmatrix} 16.4 & 0.674 & 0 & 0 & 0 & 0 \\ 0.674 & 25.2 & 0.847 & 0 & 0 & 0 \\ 0 & 0.847 & 246.5 & 0 & 0 & 0 \\ 0 & 0 & 0 & 0.240 & 0 & 0 \\ 0 & 0 & 0 & 0 & 0.173 & 0 \\ 0 & 0 & 0 & 0 & 0 & 2.58 \end{bmatrix}$$

Technical Elastic Constants

E_{11} = 16.4 GPa

E_{22} = 25.2 GPa

E_{33} = 246.4 GPa

ν_{12} = -0.000141 ν_{21} = -0.0000921

ν_{23} = 0.0336 ν_{32} = 0.00344

ν_{31} = 0.0410 ν_{13} = 0.00274

G_{12} = 2.58 GPa

G_{23} = 0.240 GPa

G_{31} = 0.173 GPa

*See Appendix A at the back of this paper for a detailed commentary on this table.

bonds include two intrachain and one interchain hydrogen bond per anydroglucose unit (see Figs. 13 and 16). The intrachain hydrogen bonds run along both sides of each chain and are shown in Fig. 16 to form (a) between the 03 oxygen of one glucose residue and the 05 oxygen of the next residue, and (b) between the 06 oxygen of one residue and the 02 oxygen of the next. The respective lengths of these hydrogen bonds, from X-ray diffraction and infrared dichroism data, are on the order of 0.275 and 0.287 nm. The interchain hydrogen bond, from 06 - H to 03 of the neighboring chain along the a axis, has a length of approximately 0.279 nm. the likelihood that an interchain 01 to 06 bond exists is discounted by these recent studies. The separation distance, 0.291 nm, is satisfactory, but the C6 - 06···01' bond angle is only 77.5°, which is too acute for an 06 - H···01' hydrogen bond to form. Accordingly, the most recent crystallographic structure studies indicate that the hydrogen bonding network is completely contained in the (020) plane and that there is no hydrogen bonding along the unit cell diagonals or along the b axis. One can thus think of the cellulose I crystal as being composed of plates of chains joined laterally by hydrogen bonds, with van der Waals contacts as the attractive forces between successive plates. If these recent structure proposals withstand further scrutiny, the calculations for elastic constants may have to be extensively revised. A Gillis-type model will have much merit for determining the two-dimensional elastic constants associated with the (020) plane (a model for 06 - H···03' links can be developed just as readily as one for 06 - H···01' links), but an entirely different approach may be needed with respect to determination of the 3-dimensional crystal.

It has been mentioned previously that the matrix is composed of the substances and spaces that surround the microfibrils. Since the distribution of the hemicelluloses, lignin, void space, etc., is non-uniform, there is point-to-point variation in the matrix elastic constants. Regions of dense molecular aggregation will have high local moduli of elasticity and rigidity, and the more open regions where air and/or water is present in the cell wall voids will have small, perhaps insignificant, moduli. The moduli overall will take on some sort of average values, but deformation will take place preferentially in the regions of lower aggregate macromolecular packing density.

It is important to note that work is required to deform the matrix molecular structure even if the matrix components are not bonded to each other or to the microfibrils. The physical location of the constituent polymer molecules between microfibrils requires that these molecules deform along with the microfibrils.

We can assume that the matrix polymers are subject to deformation via small rotations about various C - C and C - 0 bonds

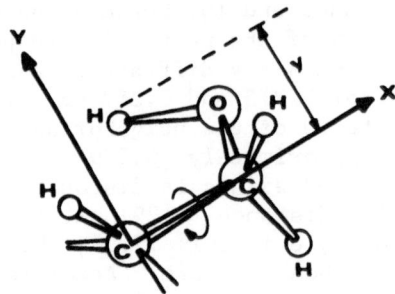

Fig. 19.   Rotation of a CH$_2$OH (hydroxymethyl) group about C — O axis.
Adpated from Bowden[6].

(see for example Fig. 19).   Such rotations require less energy than
stretching or bending of primary bonds.   Matrix moduli can be anticipated to be low since deformations of this kind (in a non–crystalline arrangement) can occur so easily.   In Fig. 19 a three-
dimensional side group is rotated about a carbon-carbon bond whose
direction is taken as the X axis.   The dimensions in the Y direction
will vary considerably with the angle of rotation.   The same can be
said of the Z direction, normal to the plane of the paper.   However,
the X dimension will stay constant.   If additional rotation occurs
in the cargon-oxygen bond, all three dimensions will change in a
manner fixed by the relative rotational movement of the C — C and
C — O bonds.   In a polymer, there are many possible conformations
because of the large number of rotatable bonds in each molecule.

      The energy barriers to intramolecular bond rotation arise
from:

      1.   Steric hindrance between substituent groups on the back-
           bone atoms.

      2.   Hydrogen bonding within the large molecule.

   3.   Other attractions between the kinked segments of nonpolar
        molecules.

   If hindered bond rotation is the controlling factor, it is possible to make an estimate of the appropriate Young's modulus for an
amorphous polymeric matrix. With reference to Fig. 19, let us
assume that a C — C bond in a matrix molecule is rotated as shown
through a small angle $\Delta\theta$. Furthermore, let us imagine that other
molecular segments in the matrix are deforming in such a manner
that the spatial domain occupied by this side chain is unchanged in
the Z direction. Since the domain in the X direction will not
change by rotation, the area A occupied by the CH_2OH group in the
X-Z plane will not change. However, a positive change in the Y direction Δy will occur, and a unit of work ΔU is accomplished by the
applied force F

   $$\Delta U = F\Delta y \simeq Fy\Delta\theta$$

where y is the domain occupied by this group in the Y direction.
Accordingly,

   $$F \simeq (\frac{1}{y})(\frac{dU}{d\theta})$$

   The stress acting on the area A will be F/A, and thus the
matrix Young's modulus E_m can be found by taking the differential
of stress, d(F/A), to strain, dy/y. That is,

   $$E_m = d(\frac{F}{A}) \div \frac{dy}{y} = d(F/A)/d\theta$$

The energy of rotation is estimated from

   $$U = P \sin^2 \phi$$

where U = energy of rotation

   P = height of energy barrier to C — C rotation ($\simeq 2.2 \times 10^{-20}$ J)

   $\phi = \frac{3}{2} \theta$, as U must be zero when θ is rotated by 120°
        (a rotation from one minimum bond energy position to
        the next)

Combination of the last two equations yields

   $$E_m = (d^2U/d\theta^2)/Ay$$

The value of dU is obtained by differentiating the term $P \sin^2 \phi$, and from this

$$\frac{d^2U}{d\theta^2} = \frac{9}{2} P(2 \cos^2 \phi - 1)$$

For small rotations, the term in parentheses becomes unity and $E_m = 4.5 \, P/Ay$. Substituting $P = 2.2 \times 10^{-20}$ J, $A = 8.4 \times 10^{-20} m^2$, $y = 0.15$ nm, we obtain 1.8 GPa. Such a value is of the same order of magnitude as experimentally obtained for lignin and other amorphous polymers. A value of 1.8 GPa for the Young's modulus E_m can be substituted into the appropriate equation for determining the shear modulus G_m of an elastically isotropic material

$$G_m = \frac{E_m}{2(1 + \nu_m)}$$

where ν_m is the Poisson ratio, here taken as 0.30. Solution of the above equation yields $G_m = 0.7$ GPa.

The actual values for E_m and G_m vary substantially with the amount of water present per unit volume of cell wall ΔV. Water acts as a plasticizer for the non-cellulosic (matrix) phase. Water expands the distances between microfibrils and reduces the physical constraints to matrix molecular deformation. Heat also reduces E_m and G_m. However, the effects of heat and moisture on wood cell walls are in general mostly reversible unless high temperatures are maintained and/or the wood is restrained from recovery, as in the case of steam bending wood over a form.

2.E. Properties of the Composite

Once the structural functions of the various chemical constituents in the wood cell wall are determined, a variety of modeling systems can be employed to determine the physical and mechanical properties of the composite unit volume element of the fiber wall. In the next section some details concerning the orientation of microfibrils in the cell wall will be provided. For now, suffice it to say that, in general, the microfibrils are not aligned with the fiber axis. Therefore, a small element of a cell wall layer, such as shown in Fig. 20, has a filament alignment that differs by some angle α from the axial direction of the fiber.

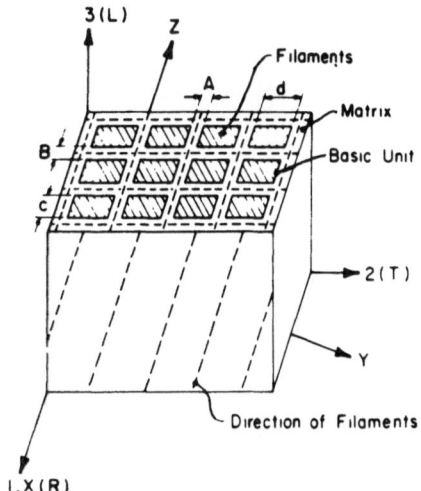

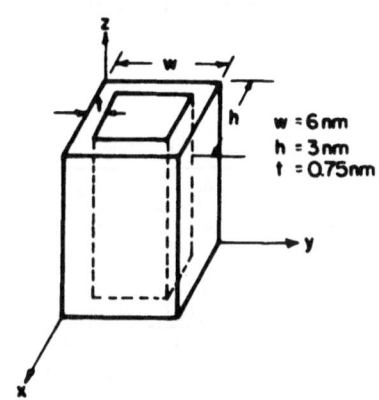

Fig. 20. Model of a submicroscopic element in a
layer of a fiber wall.
From Tang and Hsu[26].

Fig. 21. Model of a basic unit
of cell wall. The unit
contains a single rec-
tangular anisotropic
filament surrounded by
a matrix of thickness t.

The two components of a submicroscopic element of cell wall in
Fig. 20 are: the approximately rectangular reinforcing filaments
(microfibrils) and the matrix that surrounds the microfibrils.
Fig. 20 shows the microfibrils in relation to their individual
elastic coordinates (X,Y,Z) and the geometric coordinates of the
fiber wall (1,2,3). A basic unit of cell wall material, containing
a single anisotropic microfibril surrounded by isotropic matrix
material, which repeatedly produces the cell wall layer, is shown in
Fig. 21.

If one wishes to determine the material constants that flow
from elasticity theory using a given model such as shown in Figs. 20
and 21, some variation of the law of mixtures (17,18) is employed.
The following equations might be used, for example, for a two-
dimensional analysis of a thin sheet or lamina of microfibrils
composed of individual units, such as shown in Fig. 21, whose transverse elastic constants are equal for the X and Y directions in

the microfibril.

$$E_L = \frac{A_F}{A} E_{FZ} + \frac{A_M}{A} E_M$$

$$E_T = \frac{E_M E_{FT}}{E_{FT} \frac{A_M}{A} + E_M \frac{A_F}{A}}$$

$$G_{LT} = \frac{G_{FZT} G_M}{G_{FZT} \frac{A_M}{A} + G_M \frac{A_F}{A}}$$

$$\nu_{LT} = \frac{A_F}{A} \nu_{FZT} + \frac{A_M}{A} \nu_M$$

$$\nu_{TL} = \nu_{LT} \frac{E_T}{E_L}$$

where E_L, E_T = moduli of elasticity of a cell wall lamina
                        parallel and normal to the mean microfibrillar
                        direction of the lamina, respectively

          G_{LT} = shear modulus of rigidity of the lamina in the
                     plane of the lamina

          ν_{LT}, ν_{TL} = Poisson ratios for a cell wall lamina, giving
                        ratios for contraction in the direction indicated
                        by the second subscript to extension under stress
                        in the direction indicated by the first sub-
                        script, in the plane of the lamina

          $\frac{A_F}{A}$, $\frac{A_M}{A}$ = proportions of a unit element of cell wall in
                        microfibril (framework) and matrix, respectively

          E_{FZ} = modululus of elasticity of microfibrils in the
                     chain axis direction

E_{FT} = modulus of elasticity of microfibrils in the
     direction normal (transverse) to the chain axes

E_M = matrix modulus of elasticity

G_{FZT} = shear modulus of rigidity of the microfibril in
     a ZT plane

G_M = shear modulus of rigidity of the matrix

ν_{FZT}, ν_M = appropriate Poisson ratios for the microfibrils
     and matrix, using same subscript convention as
     above

Similar, but more complex, equations are used for the full
three-dimensional determination of cell wall elastic constants. One
may refer to several recent papers for these formulations[4,15,22,26].

The above equations, when used with appropriate data such as
an assumption that $\frac{A_F}{A} = \frac{A_M}{A} = 0.5$, the values for matrix constants
given at the end of the preceding section, and the values for cellu-
lose I (framework) in Table 1, yield values for elastic constants
on the order of those shown below:

E_L = 124 GPa

E_T = 3.4 GPa

G_{LT} = 0.36 GPa

ν_{LT} = 0.15

ν_{TL} = 0.004

The model shown in Figs. 20 and 21 may be modified somewhat if
it seems more appropriate. An example appears in Fig. 22, used by
Norimoto et al.[21] to determine the dielectric constants for wood
cell wall substance. In this model, there is a 3-phase structure
consisting of, 1st, crystalline cellulose, designated as "c" in
Fig. 22; 2nd, mannan-containing hemicelluloses plus "noncrystalline"
cellulose, shown as "n" in Fig. 22; and 3rd, other cell wall components, designated as matrix (m).

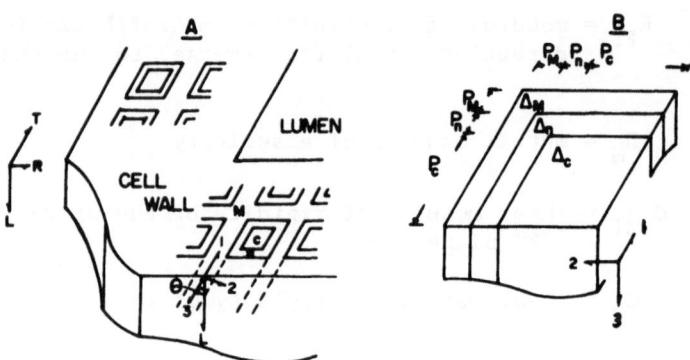

Fig. 22. Cell wall model used by Norimoto et al.[21] to calculate anistropic dielectric constants.

The dieletric constants in the 1-, 2-, and 3-directions of cellulose I may be estimated from

$$\varepsilon_{c1} = \varepsilon_{c2} \simeq \frac{1.59}{1.55} (\varepsilon_{002} - 1) + 1$$

$$\varepsilon_{c3} \simeq \frac{1.59}{1.55} (\varepsilon_{003} - 1) + 1$$

where ε_{002} and ε_{003} are the intrinsic dielectric constants at infrared frequencies perpendicular and parallel to the chain axis direction.

The calculated values are 4.01, 4.01, and 4.25, respectively.

For the "noncrystalline" and matrix phases, Norimoto et al.[21] have constructed a graph, reproduced in Fig. 23, which shows these constants to be generally heavily dependent upon the moisture content of the cell wall, as expected, while the curves for crystalline cellulose are unaffected by changes in cell wall moisture.

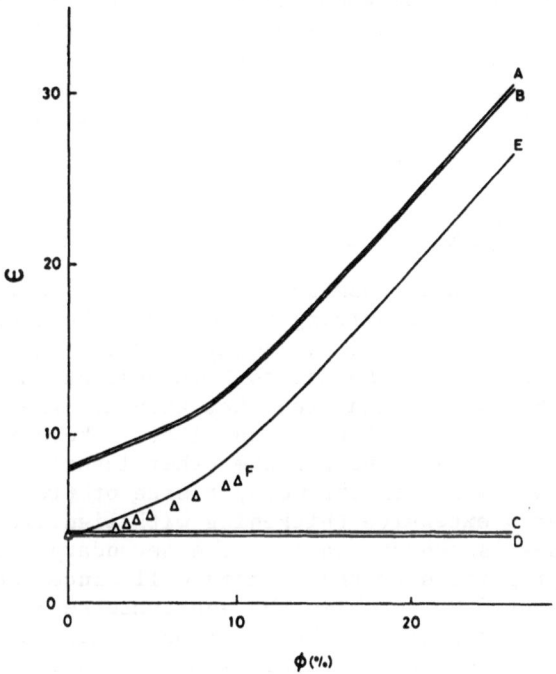

Fig. 23. Intrinsic dielectric constants (ordinate) vs. moisture
content (abscissa) at 20°C and 1 MHz. Symbols represent

A. Non-crystalline substance parallel to chain direction
    (constant taken to be equal to that for mannan).

B. Mannan (also other non-crystalline substance) normal
    to chain direction.

C. Crystalline substance parallel to chain direction
    (constant taken equal to that for cellulose).

D. Cellulose normal to chain direction.

E. Pentosan (e.g., xylan).

F. Experimental values for wall substance in direction of
    S2 microfibrils.

3. THE ARCHITECTURE OF A WOOD FIBER

It was mentioned, in the opening paragraph of Part 2, that a simple way to envision the wood fiber as a composite is to consider the microfibrils as loops in a skein of yarn that has been (a) pulled and twisted and then (b) embedded in a rigid matrix. At this time it is important to add some detail to that simple picture in order to achieve full comprehension of the properties of a cell in wood.

3.A. The layered Structure of Fibers and Other Cells; Assemblage

Wood fibers normally possess three distinct layers that are readily visible in the polarizing microscope; the layers are also detected by a number of chemical, X-ray, and optical and electron microscopical techniques. Botanists have designated these layers as S1, S2, and S3, which signifies that they are the 3 layers formed in the secondary (S) wall of the wood fiber. There is also a primary (P) wall layer that envelops the other layers. All plant cells have a primary wall, but in the woody tissue of trees, only those cells that undergo extensive thickening with lignification are considered to possess secondary walls. The secondary wall layers are very thick in proportion to the primary wall since the latter has a thickness of about 0.06 μm while the aggregate secondary wall thickness may be two orders of magnitude (or more) greater. The outermost (first-formed) secondary layer is S1, which is followed ontogenetically by S2 and S3, as shown in Fig. 24. The completion of the S3 layer is normally followed by the rapid death of the protoplast. The middle lamella (M), also shown in Fig. 24, provides the intercellular bond that joins the fibers and other cells in wood together. Pulping is the industrial process that separates fibers from one another. When done chemically, the fibers are separated as a result of partial or complete dissolution of the middle lamella.

Let us now revert to the concept of a skein of helically twisted yarn shown in Fig. 4 as a model for microfibrillar orientation. In our real wood fiber there are successive microfibrillar helices in the different layers whose helical angles vary markedly from one layer to another. The thickest layer, S2, tends to have a relatively consistent helical pitch for the microfibrils, but the S1 layer has a crossed helical structure due to the presence of two counter-rotating microfibrillar orientations. The S3 may in some cases also possess two or more principal orientations. The P layer has much more dispersion of its microfibrils. The reader may refer to Figs. 11 and 25-27 for diagrammatic and electron micrographic views of various microfibrillar orientations in plant cell walls. Cellulosic cell walls are characterized by helically wound laminae of microfibrils. As mentioned, the wood fiber normally

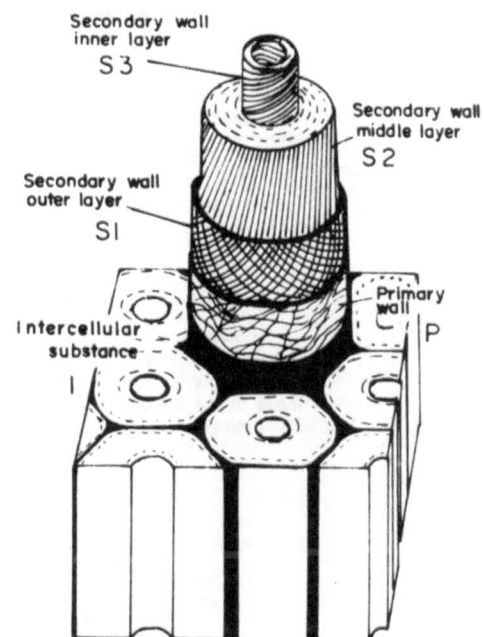

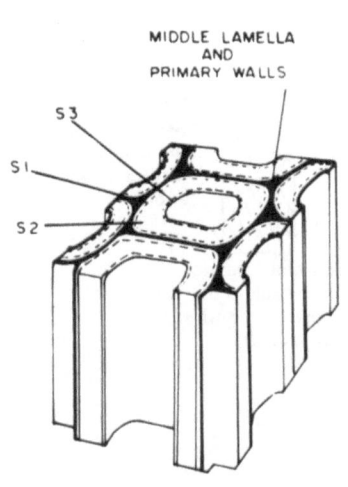

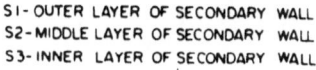

S1- OUTER LAYER OF SECONDARY WALL
S2- MIDDLE LAYER OF SECONDARY WALL
S3- INNER LAYER OF SECONDARY WALL

Fig. 24. Diagrammatic represen- Fig. 25. Diagrammatic represen-
          tation of a segment of   tation of dominant
          a conifer tracheid sur-   helical patterns in
          rounded by 6 other   layers of a wood fiber.
          tracheids, showing pri-
          mary and secondary walls.

has 3 principal secondary layers,each composed of a series of thin
sheets, or laminae. Other cell types might have a greater or lesser
number of layers (examples: in some species of spruce, S3 may be
absent or vanishingly thin; bamboos often have more than 3 layers),
but the pattern of helically-arranged microfibrillar structure is
about universal in plants. The electron microscope has revealed
that there are often thin transition layers of intermediate orientation as one passes between two principal layers (see for example
Fig. 26A).

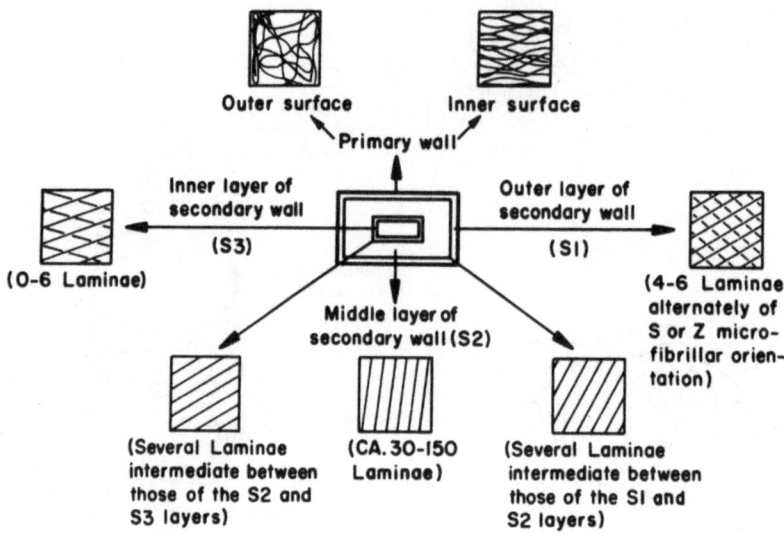

A

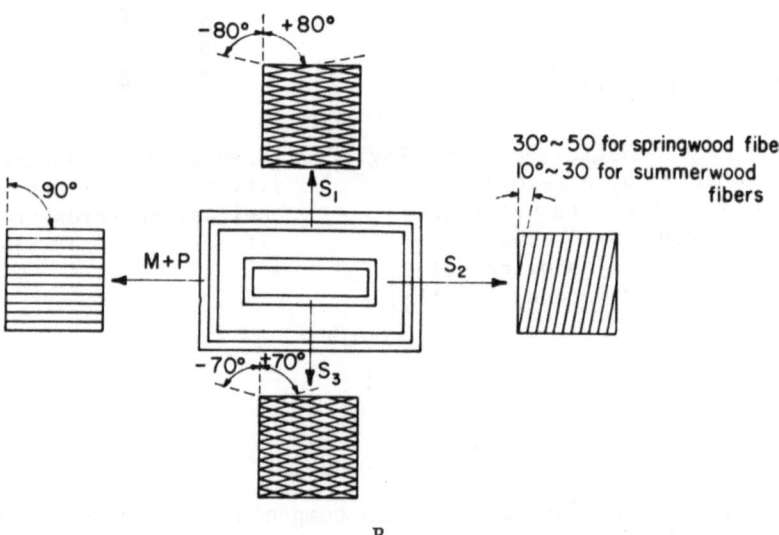

B

Fig. 26.   Schematic diagrams of the layer structure of a wood fiber.
(A) A typical structural breakdown for the layers and
transition areas between layers, showing the microfibrillar
textures encountered.   (B) A simplified version, useful
for modeling the cell wall as a laminated composite material (from Refs. 27,26).

3B. Anisotropy of the Fiber

Although the cellulose crystal is monoclinic, the space lattice is close enough to a Bravais 10 orthorhombic system that we can assume orthotropic elastic properties for the microfibrils, as was done in Section 2E for a 2-dimensional case. When the elastic constants and the volume fractions of framework and matrix components are combined as in 2E, the orthotropic material constants for a lamina, and by extension for a cell wall layer such as P, S1, S2 or S3, can be developed.

In the intact wood fiber, it must be assumed that the wall layers, shown in an exploded view in Fig. 27, will all deform together without relative slippage under applied load.

For each wall layer in the absence of swelling, Hooke's law for the microfibrillar filaments can be written as:

$$\sigma^f = C^f \, \varepsilon^f$$

Here σ denotes the stress tensor, ε the strain tensor, C the elastic stiffness tensor, and the superscript f refers to the cellulosic microfibrils or filaments. For the matrix material we use superscript m and note that arbitrary swelling strains must be provided for in Hooke's law. By analogy with thermal strains, this can be written as:

$$\sigma^m = C^m \, (\varepsilon^m - \varepsilon^*)$$

where ε^* are the swelling strains. By convention, positive strains are taken to be tensile, i.e., swelling strains, although we may refer to strains as either shrinkage or swelling, it being understood that negative strains are required to describe shrinkage.

Because $C^m \varepsilon^*$ defines a stress tensor, denoted as σ^*, the preceding equation can be written equivalently as:

$$\sigma^m = C^m \, \varepsilon^m - \sigma^*$$

The overall behavior of the composite is related to its constitutents through the principle of volume average properties. Let f be the volume fraction of filament and $m = 1 - f$ that of matrix. Then the overall stress tensor σ is assumed to be the volume average: $\sigma = f \, \sigma^f + m \, \sigma^m$. Therefore

$$\sigma = f \, C^f \, \varepsilon^f + m \, C^m \, \varepsilon^m - m \, \sigma^*$$

Thus, if the composite stiffness tensor, C, is also defined by volume averaging, $C\varepsilon = C \, (f\varepsilon^f + m \, \varepsilon^m)$, Hooke's law for the composite

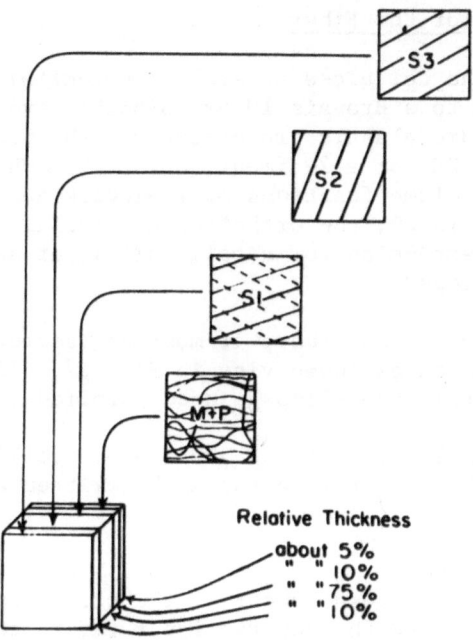

Fig. 27.   Wall layers of wood fiber cell wall, showing microfibri-
lar orientations and relative thicknesses of layers.
From Gillis and Mark[15].

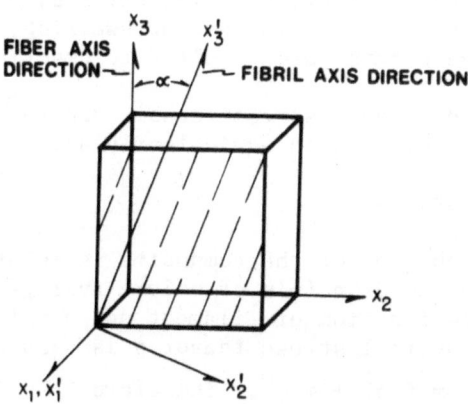

Fig. 28.   Model element of a layer of the fiber wall, showing the
reference frames with respect to microfibrillar orientation (primed axes) and the fiber direction (unprimed
axes)[15].

layer must be of the form:

$$\sigma = C \ \varepsilon \ - m \ \sigma*$$

The fiber as a whole is treated as an axially symmetric tube
consisting of several layers. The contribution of each layer to
the properties of the fiber is proportional to the relative volume
it occupies in the fiber wall. In brief, the procedure is to use
an expanded form of the preceding equation for each layer, yielding
stiffness constants that are orthotropic with respect to the reference frame of the layer, i.e., parallel and perpendicular to the
predominant microfibril direction of that layer (one perpendicular
axis lies in the plane of the wall layer; the other is oriented
radially - in the thickness direction of the layer). Transformation equations are then used to obtain stiffness constants with
respect to fiber coordinates (unprimed axes in Fig. 28) from the
stiffness constants with respect to layer coordinates (primed axes
in Fig. 28). These transformations are performed for each layer by
accounting for the angle α between the mean microfibrillar direction
of the layer and the fiber axis.

Within a layer, the normal stiffness constant C_{33} associated
with the X_3 direction will be reduced in comparison with that for
the X_3' direction (C_{33}'). As examples, if the angle α in Fig. 28
is 15°, C_{33} for the layer will be about 15 per cent less than C_{33}'.
If $\alpha = 30°$, the reduction will amount to approximately 45 per cent.

The properties of the composite fiber, composed of several
layers, are thus influenced heavily by the degree of microfibrillar
orientation and the proportions of the cell wall in each layer.

In Fig. 29 is shown the calculated relationship between the
axial Young's modulus of a wood fiber and the S2 angle (the mean
microfibrillar angle of the S2 layer) for four different S1 angles.
The assumptions used in generating these curves were that

a) The proportions of wood substance in each layer were as
    follows: primary wall and adjoining intercellular material
    (M + P), 11.17 per cent; S1 layer, 17.52 per cent; S2
    layer, 61.11 per cent; S3 layer, 10.20 per cent. These
    values were taken from actual measurements of conifer
    tracheids.*

---

*It may be noted that, although the S2 layer dominates, one cannot
 ignore the intercellular substance and other layers in most computations without introducing significant errors. Nearly two-fifths
 of wood substance lies outside S2.

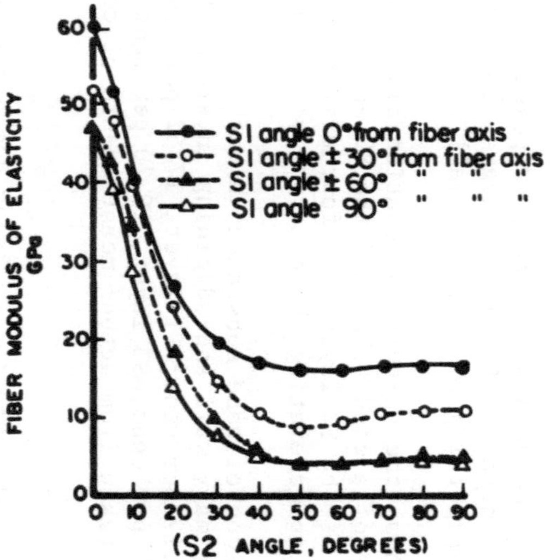

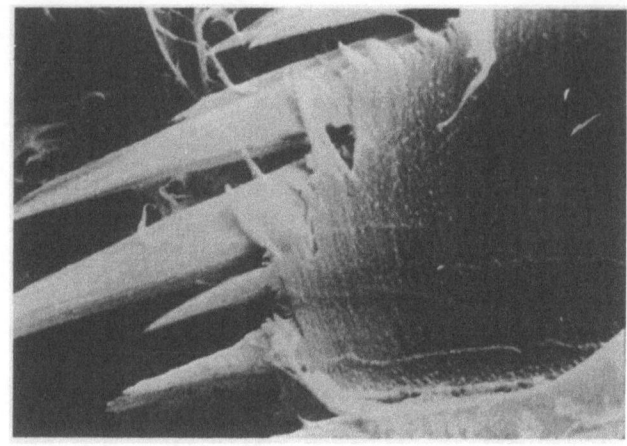

Fig. 30

Scanning electron microscopic view of a tracheid fractured in tension. The fiber wall is viewed from the lumen; in front is the S3 layer, while jagged splinters of S2 jut out behind it. The morphological differences in the fracture surface reflect the distinct microfibrillar texture and orientation differences between the two layers. Courtesy of Prof. Hiroshi Harada, Kyoto University.

Fig. 29

Theoretical curves for axial elastic modulus of a fiber in wood versus assumed S2 angle[20]. Parametric curves are for four different S1 angles. (Note: in actuality, S2 angles rarely if ever exceed 50°; an S1 angle of 0° is unknown. Typically, S2 might have an angle of 15–20°, and S1 might exhibit angles of +70° and −70°.)

b) The proportions of framework and matrix in each layer were as follows: M + P, 10.1 per cent framework and 89.9 per cent matrix; all other layers, 53.1 per cent framework and 46.9 per cent matrix. Again, values are from experimental measurements.

c) The elastic constants for matrix and reinforcement are those given in Section 2.D.

Many refinements of this basic method have been developed for various applications in problems in wood, fiber, paper and cellulose science. For example, one can, using appropriate boundary conditions and equations of equilibrium and compatibility, predict the locations of concentrated stress in cell walls under elastic deformation conditions and extend this analysis to large deformations and even failure with appropriate modification.. In spite of many chemical, structural, and geometric factors that might lead us to expect non-Hookean behavior, wood normally exhibits a highly linear stress-strain relationship under load*, which tells us that cellulose I itself <u>in the microfibril</u> obeys Hooke's law very closely and that its stiffness is so great in relation to the associated polymers that its influence vastly predominates over the effects of cellular geometry and nonstructural molecular components. Another use of cell wall models is in predicting shrinkage and swelling behavior[2,3].

The implications of fiber cell wall architecture are varied. Fracture, for example, is very different morphologically in the S1 or S3 layers than it appears in the S2 (see Fig. 30). Fibers that have been isolated from wood undergo extensive twisting upon drying unless they are restrained in some manner (e.g., as in formation of paper where fibers hydrogen-bond to one another and create mutual restraint) because the dominant S2 layer has an unbalanced helical reinforcement of microfibrils. Note, however, that the adjacent cell walls <u>in wood</u> tend to create a <u>balanced</u> laminate through the double-wall thickness. Adjacent cell walls not only tend to balance each other with respect to microfibrillar orientation, but also tend to be of similar thickness, as shown in Fig. 3. Thin-walled fibers that have been separated by chemical pulping undergo extensive collapse and flattening in the sheet (see Fig. 31). Thus, the large-lumen, thin-walled, relatively weak fibers characteristic of low-density wood make paper of high density and generally favorable mechanical properties. Also, bonding of the fibers in paper is enhanced by "beating", a process which involves the roughening of the S1 layer, which creates more surface area for hydrogen bonding between crossing (overlapping) fibers.

---

*Exceptions: Loads imposed for long durations, conditions of large ambient humidity changes.

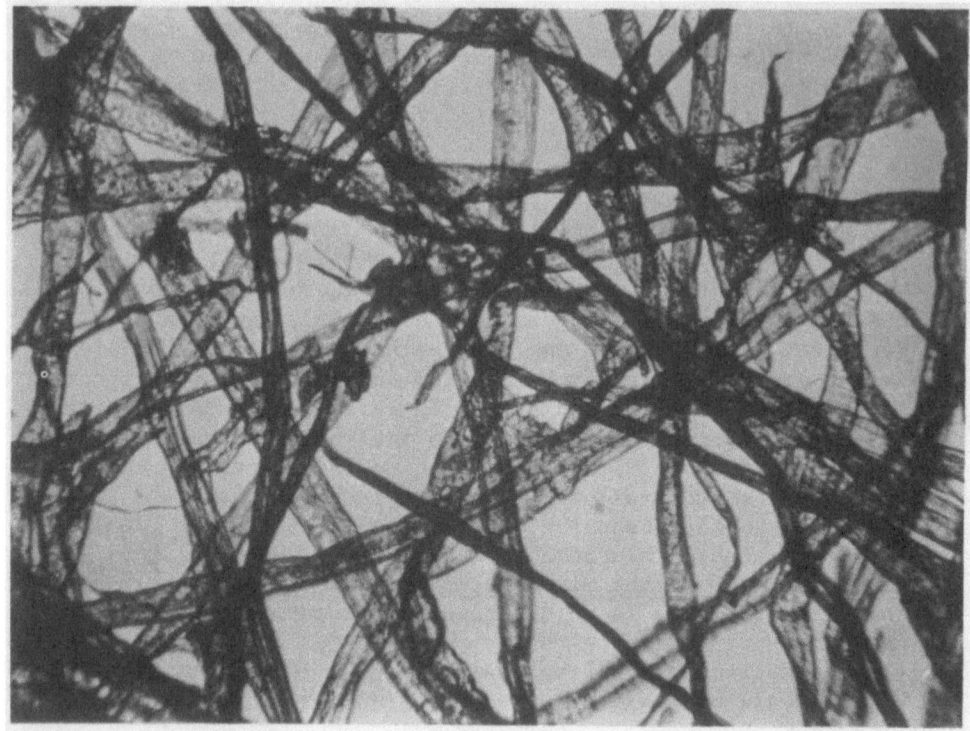

Fig. 31. Chemically pulped fibers exhibit a collapsed, flattened
          appearance as a result of the partial removal of matrix
          materials during the pulping. Courtesy Empire State
          Paper Research Institute, SUNY College of Environmental
          Science and Forestry, Syracuse.

    The process of beating or refining fibers does more than
roughen or "fibrillate" the surface. It also causes a stretching
of the fibers. Then, since a steep helical spring with a small
angle between axis and coil directions will shrink more in the
radial direction than a flatter helix with a large coil angle, it
follows that internal helicoidal breaks will occur, first at the
S1-S2 interface and later within S2 itself. These internal sepa-
rations of the lamellae of microfibrils within the S2 layer raise
the hydrodynamic specific surface from about 1.2 m^2/g to about
4 m^2/g. When the fibers are formed into sheets of paper and dried,
these lamellae rejoin by hydrogen bonds but in a new, flatter,
denser configuration.

APPENDIX A

(Footnote to Table 1)

*Although a 4-index notation is sometimes employed for elastic con-stants, an abbreviation to a 2-index system may be employed for sim-pler operation. The simplifications are made as follows for stresses:

$$\sigma_{11} = \sigma_1 \qquad\qquad \sigma_{23} = \sigma_4$$
$$\sigma_{22} = \sigma_2 \qquad\qquad \sigma_{13} = \sigma_5$$
$$\sigma_{33} = \sigma_3 \qquad\qquad \sigma_{12} = \sigma_6$$

and similarly for strains

$$\varepsilon_{11} = \varepsilon_1 \qquad\qquad 2\varepsilon_{23} = \varepsilon_4$$
$$\varepsilon_{22} = \varepsilon_2 \qquad\qquad 2\varepsilon_{13} = \varepsilon_5$$
$$\varepsilon_{33} = \varepsilon_3 \qquad\qquad 2\varepsilon_{12} = \varepsilon_6$$

Thus, Hooke's law becomes

$$\sigma_i = \sum_{j=1}^{6} S_{ij}\, \varepsilon_j \quad ,$$

where the S_{ij} are the compliance moduli. Accordingly, the matrix of stiffness moduli C_{ij} would be (for a monoclinic crystal):

C_{11}	C_{12}	C_{13}	0	0	C_{16}
	C_{22}	C_{23}	0	0	C_{26}
		C_{33}	0	0	C_{36}
			C_{44}	C_{45}	0
				C_{55}	0
					C_{66}

yielding 13 independent constants (with reference to orthogonal axes) For an (assumed) orthorhombic unit cell, 9 constants remain, viz.:

C_{11}	C_{12}	C_{13}	0	0	0
	C_{22}	C_{23}	0	0	0
		C_{33}	0	0	0
			C_{44}	0	0
				C_{55}	0
					C_{66}

REFERENCES

1. Adler, E. 1977. Lignin chemistry - past, present and future.
        Wood Sci. Technol. 11:169-218.
2. Barber, N. F. 1969. The shrinkage of wood, theoretical models.
        I.P.P.S. Conference on Science of Materials, Auckland,
        N. Z. Inf. Ser. 71, NZDSIR 184.
3. Barber, N. F. and Meylan, B. A. 1964.   The anisotropic shrink-
        age of wood, a theoretical model.   Holzforsch. 18: 146-156.
4. Barrett, J. D., and Schniewind, A. P. 1973. Three-dimensional
        finite-element models of cylindrical wood fibers. Wood and
        Fiber 5(3):215-225.
5. Blackwell, J., and Kolpak, F. J. 1975. The cellulose micro-
        fibril as an imperfect array of elementary fibrils. Macro-
        molecules 8:322-326.
6. Bowden, P. B. 1968. The elastic modulus of an amorphous
        glassy polymer. Polymer 9(9):449-454.
7. Browning, B. L. 1963. The composition and chemical reactions
        of wood. in "The Chemistry of Wood," B. L. Browning, Ed.,
        Interscience, New York 57-101.
8. Bucher, H. 1958. Discontinuities in the microscopic structure
        of wood fibres, in "Fundamentals of Papermaking Fibers,"
        F. Bolam, Ed., Brit. Pap. Bd. Assoc., Surrey 7-26.
9. Caulfield, D. F. 1971.   sizes in wet and dry
        Valonia ventricosa. Textile Res. J. 41:267.
10. Fergus, B. J. and Goring, D. A. I. 1970. The distribution of
        lignin in birch wood as determined by ultraviolet micro-
        scopy. Holzforsch. 24(4):118-124.
11. Frey-Wyssling, A. and Mühlethaler, K. 1965. "Ultrastructural
        Plant Cytology," Elsevier Pub. Co., New York.
12. Frey-Wyssling, A., Mühlethaler, K. and Muggli, R. 1966. Ele-
        mentarfibrillen als Grundbausteine der nativen Cellulose.
        Holz Roh-Werkstoff 24:443-444.
13. Gardner, K. H., and Blackwell, J. 1974. The structure of
        native cellulose. Biopolymers 13:1975-2001.
14. Gillis, P. P. 1969. Effect of hydrogen bonds on the axial
        stiffness of crystalline native cellulose. J. Polymer Sci.
        A2 7:783-794.
15. Gillis, P. P. and Mark, R. E. 1973. Analysis of shrinkage,
        swelling and twisting of pulp fibers. Cellulose Chem. Tech.
        7(2):209-234.
16. Goring, D. A. I. 1971. Polymer properties of lignin and
        lignin derivatives, in: "Lignins, Occurrence, Formation,
        Structure and Reactions," K. V. Sarkanen and C. H. Ludwig,
        eds, Wiley-Interscience, New York 695-768.
17. Hill, R. 1963. Elastic properties of reinforced solids:
        some theoretical principles. J. Mech. Phys. Solids 11:
        357-372.

18. Hill, R. 1964. Theory of mechanical properties of fibre-
    strengthened materials. I. Elastic behaviour. $\underline{J}$. $\underline{Mech}$.
    $\underline{Phys}$. $\underline{Solids}$ $\underline{12}$:199–212.

19. Jaswon, M. A., Gillis, P. P., and Mark, R. E. 1968. The elas-
    tic constants of crystalline native cellulose. $\underline{Proc}$. $\underline{Roy}$.
    $\underline{Soc}$. (London) $\underline{A}$ $\underline{306}$:389–412.

20. Mark. R. E. 1972. Mechanical behavior of the molecular com-
    ponents of fibers, $\underline{in}$: "Theory and Design of Wood and Fiber
    Composite Materials," B. A. Jayne, ed., Syracuse Univ.
    Press 49–82.

21. Norimoto, M., Hayashi, S., and Yamada, T., 1978. Anisotropy
    of dielectric constant in coniferous wood. $\underline{Holzforsch}$. $\underline{32}$
    (5):167–172.

22. Ohgama, T., Masuda, M., and Yamada, T. 1977. Stress distri-
    bution within cell wall of wood subjected to tensile force
    in transverse direction (in Japanese). $\underline{J}$. $\underline{Soc}$. $\underline{Materials}$
    $\underline{Sci}$. (Japan) $\underline{26}$:433–438.

23. Preston, R. D. 1974. "The Physical Biology of Plant Cell
    Walls," Chapman and Hall, London.

24. Sarkanan, K. V., and Ludwig, C. H. (eds.). 1971. "Lignins:
    Occurrence, Formation, Structure and Reactions," Wiley-
    Interscience, New York.

25. Sarko, A., and Muggli, R. 1974. Packing analysis of carbo-
    hydrates and polysaccharides. III. $\underline{Valonia}$ cellulose and
    cellulose II^{1a}. $\underline{Macromolecules}$ $\underline{7}$:486–494.

26. Tang, R. C., and Hsu, N. N. 1973. Analysis of the relation-
    ship between microstructure and elastic properties of the
    cell wall. $\underline{Wood}$ $\underline{\&}$ $\underline{Fiber}$ $\underline{5}$:139–151.

27. Wardrop, A. B. 1964. The structure and formation of the cell
    wall in xylem, $\underline{in}$: The Formation of Wood in Forest Trees,"
    M. H. Zimmerman, ed., Academic Press, N. Y. 87–134.

28. Woodcock, Carrie. 1979. "The X-ray crystallographic analysis
    of the structure of native ramie cellulose," M. S. Thesis,
    SUNY College of Environmental Science and Forestry, Syracuse,
    New York.

STRUCTURAL WOOD ADHESIVES -

TODAY AND TOMORROW

Roland E. Kreibich

Chemistry Department
Weyerhaeuser Technology Center
Tacoma WA 98477

INTRODUCTION

Historically the wood products industry has supplied rods and sheets for building structures. To the customer we offer a structural element with a favorable strength to weight ratio or stiffness to weight ratio.

By the use of adhesives, natural or synthetic, we were able to deliver practically any size needed or wanted. With the development of wood adhesives based on phenol (and its homologues) the durability of properly manufactured bonds was no longer a problem. Thus, phenolic type adhesives have dominated the fully durable exterior area for about one-half century. During this same time macrosections of high quality raw material were readily available.

We may now have reached the turning point in both areas: Our high quality, large size cellulosic substrate has already disappeared. Benzene, nature's stable nucleus and the foundation for wood adhesive durability, sometimes called the grandfather of the structural bond, is in high demand for other uses, is at times in short supply, and has become relatively expensive.

The path, then, is clear: Wood Technology must learn to produce small sections of specified geometry and critically evaluate resulting product performance as dependent on geometry. Chemistry must learn to replace the benzene ring, in part or in total, with other, more easily accessible compounds without sacrificing bond quality or bond durability, at least not to a greater degree than the end product can tolerate.

Durability

Durability is the capability of a bond to endure bond degrading factors to which the product is exposed under normally anticipated use conditions.

In my general and daily discussions with individuals concerned with wood adhesion, I find uncertainty and disagreement regarding the terms "quality" and "durability." Referring to Fig. 1, the indicator for bond line quality such as (common) shear strength or percent wood failure is plotted on the ordinate. On the abscissa is plotted the severity, duration or cycles of bond degradation, such as time of exposure, number of cycles. I propose we regard Adhesive A as having high quality and high durability, Adhesive B as having high quality and low durability, Adhesive C as having low quality and high durability. (I have assumed that during our bond line degradation process the substrate remains essentially unchanged.)

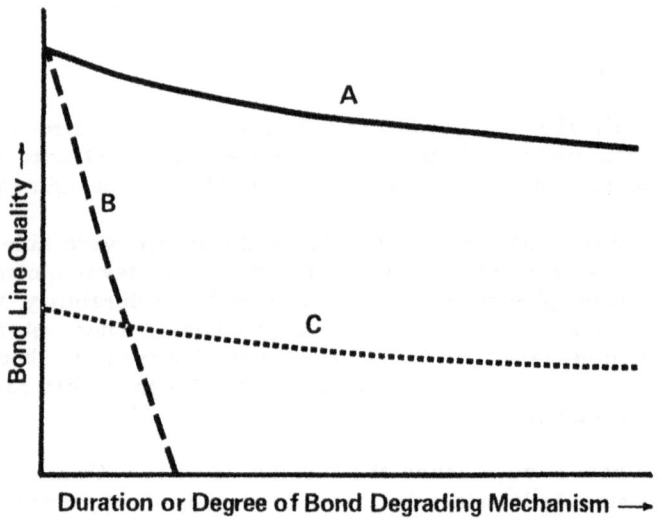

Fig. 1. Distinguishing Adhesive Quality and Durability

For structural purposes our key interest of course lies in the accurate prediction of bond line durability.

In our field of wood adhesives, most durability prediction methods involve changes in both moisture content and temperature of bond line and substrate.

There are exceptions to this, however. One method developed is based on the Arrhenius theory and subjects specimens to elevated temperature exclusively. The target is to establish a correlation of temperature and half life of the bond.

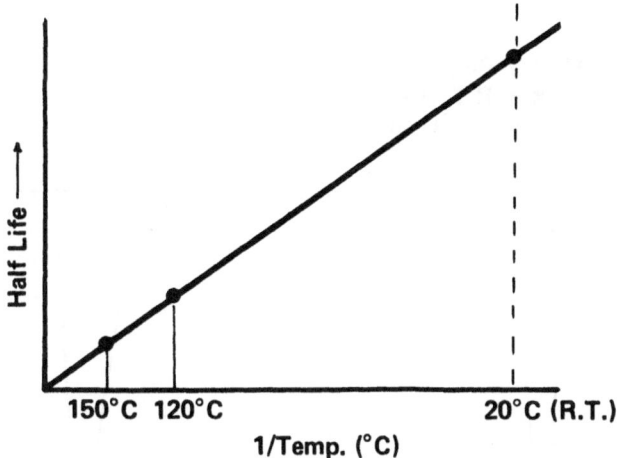

Fig. 2.   Correlation of Temperature and Half Life

Realizing that the wood product during its life span may experience a great variety of exposure conditions, we prefer to design (or over-design) bond line quality and durability on the conservative side. In parallel, the test methods or the criteria used during the evaluation of the test data often tend to be in the conservative direction.

After personal involvement in claims as well as durability prediction test development for the industry, I too have accepted a conservative attitude. This was one of the driving forces for the development of an automatic boil/dry system which subjects specimens to any desired number of boil/dry cycles without the need of manual labor. (See Fig. 3.) Details are given in ASTM D-3434-75.

When exposing different adhesive types to such methods, one finds that adhesive systems based on different chemistry may respond differently. Some will fall off to begin with, then level out. Others will maintain a good performance level for some time, then fall off sharply after many cycles. A third group may show a nearly linear degradation. For typical examples see Fig. 4.

After testing dozens of different adhesive types with this method, some general statements can be made:

a.   All properly processed phenolic types maintain highest performance up to 1600 cycles of boiling and drying. So far, no other adhesive type has been found to be equal in durability to the phenolic type.

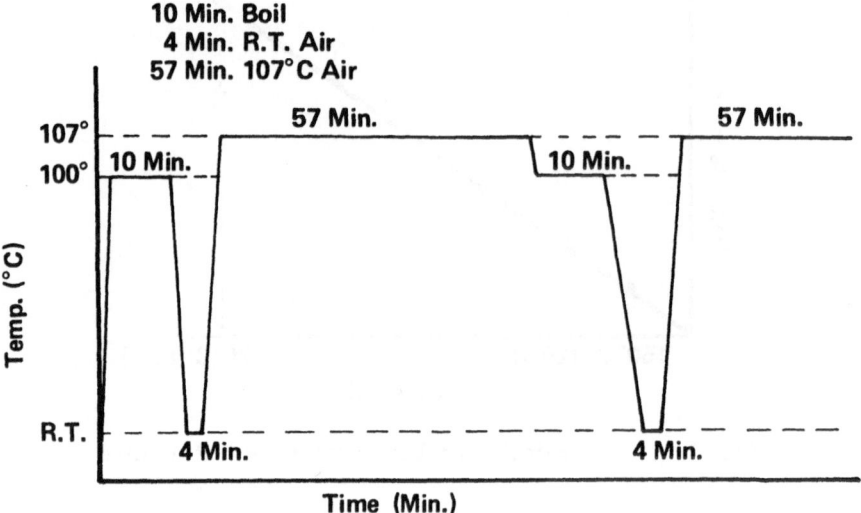

10 Min. Boil
4 Min. R.T. Air
57 Min. 107°C Air

Fig. 3 .   Cycle of Automated Boil/Dry Test

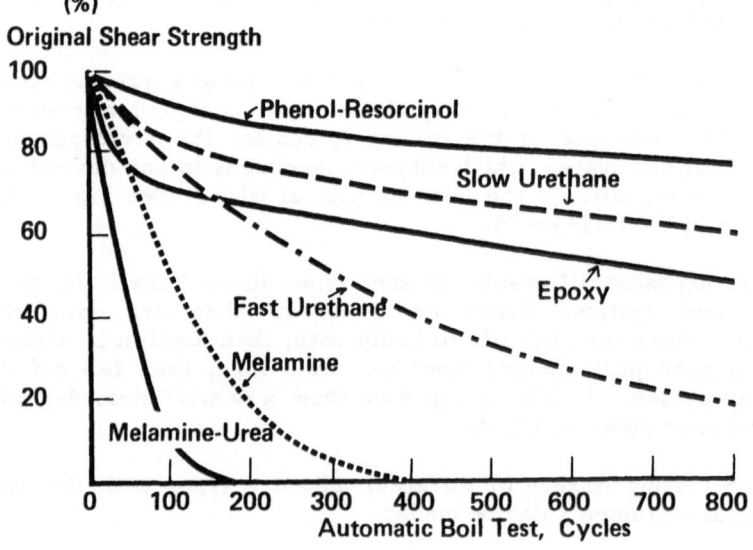

Fig. 4.   Typical Durability Curves

b. We have never seen a urea based system that was not completely degraded (full delamination) within one day (20 cycles).

c. The epoxies, polyurethanes, melamines, etc., lie anywhere in the spectrum between the phenolics and the ureas. They vary greatly depending on manufacturer, composition, type, etc.

Overall, this method has the advantages of speed, relative low cost, processing large numbers of samples, and good reproducibility. It may have the disadvantage of being very severe and may therefore tend to favor adhesive systems over-designed for a given application. This, of course, becomes much a matter of philosophy and opinion.

Test Methods (In General)

I like to distinguish three major categories:

Methods to determine if 1) the adhesive is able to produce a bond of sufficient strength, 2) the bond has durability, and 3) the adhesive allows creep.

Methods to determine if the adhesive is able to produce a bond of sufficient strength. A great variety of methods is used for this purpose. Most involve some accelerated aging followed by mechanical destruction.

For production line adhesives with unusually rapid cure it can be important to determine the initial strength-time relationship. For this purpose, a double acting pneumatic system can be used. (See Fig. 5.)

This apparatus was designed for cross lap specimens with the compression and the application of tension accomplished by a pneumatic (reversing) system.

Methods to determine if the bond (regardless of strength level) has durability. This area has received a great deal of attention from individuals as well as from agencies and corporations. For the manufacturer of structural building elements (especially laminated beams, posts, trusses or plywood used in structural applications) an exact knowledge of bond line durability is highly desirable. Claims from failures of fully structural elements can easily wipe out entire years' profits and have in the past occasionally wiped out entire corporations. One of the preferred methods for testing durability was described previously.

Methods to determine if the adhesive shows (or allows) creep. This field of adhesive evaluation has received relatively little attention in the wood industry. In comparison to the determination of bond strength and bond durability not many methods for creep have been explored, developed or published. Some of these, originally developed for metal-to-metal bonding, have been adapted for wood substrates. Many use a stack or series of specimens with the glue lines under the desired load. In order to change the

environment or to cycle the specimens through a prescribed humidity/
temperature/time sequence, the stack has to be disassembled and rebuilt.
For a multi-cycle exposure, this not only represents a substantial manpower
effort but also raises questions about the reproducibility.

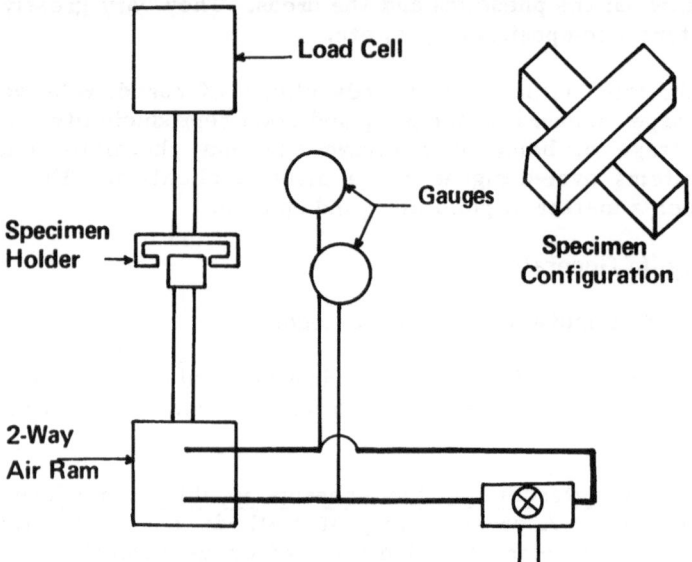

Fig. 5. Pneumatic Tester

To this date creep has played such a minor role in test development
because the natural as well as the synthetic wood adhesives used have been
relatively free of this problem. This may change in the foreseeable future, as
more adhesive types of thermoplastic origin (backbone) with little understood
or marginally adjustable cross-linking mechanisms are introduced to our
industry. I am expecting the development in this direction to intensify over
the next decade, and with it we should anticipate an intensification of all
phases of the creep phenomenon, including studies of a theoretical nature,
laboratory experimentation to compliment the theory, efforts to implement
these findings, and finally development of a series of practical test methods.
I believe this is one field of exploration that must receive a fair share of
attention before a wide variety of new non-oil-based polymers will receive
serious consideration and acceptance in the field of structural wood adhesives. Prior to the development of meaningful test methods our production
facilities will justifiably hesitate to accept new and substantially different
systems.

Chemistry

Within the scope of this paper, it will be possible to point out only the
most fundamental aspects of very few systems. (For details, the corresponding literature must be consulted.)

Phenolics.   This includes phenol, resorcinol, m-hydroxyaniline, m-phenylenediamine, phloroglucinol, and cresols, in reaction with formaldehyde.

The formaldehyde adds to one or more of the reactive positions of the phenol forming a methylol (or hydroxymethyl) derivative, which on further reaction (depending on pH, temperature, type of catalyst) may undergo polymer formation through the stages of methylene ethers and/or methylene bridges.   See Fig. 6 for a graphic representation of this sequence.

Aminoplasts.   Urea and melamine condensation products with aldehydes will undergo analogous reactions as shown in Figure 6.

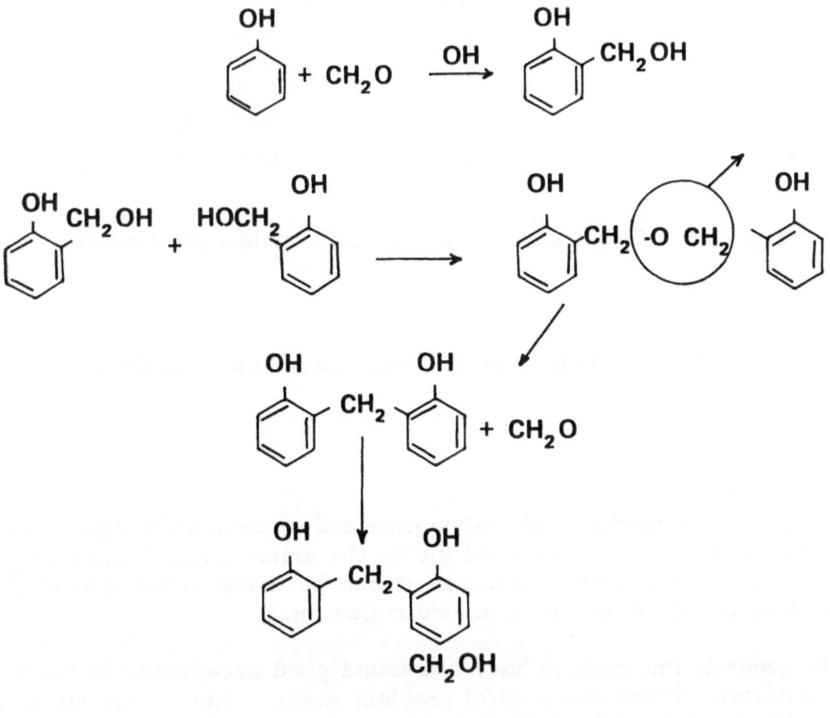

Fig. 6a.   Reaction of Phenol with Formaldehyde (Alkaline).

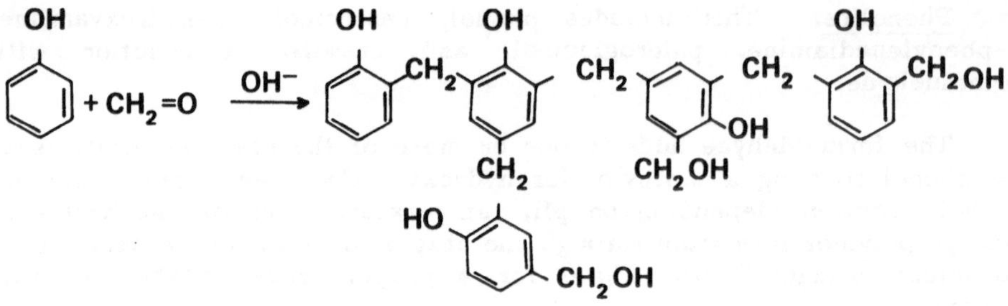

Fig. 6b.   Preparation of a Phenol–Formaldehyde Resole

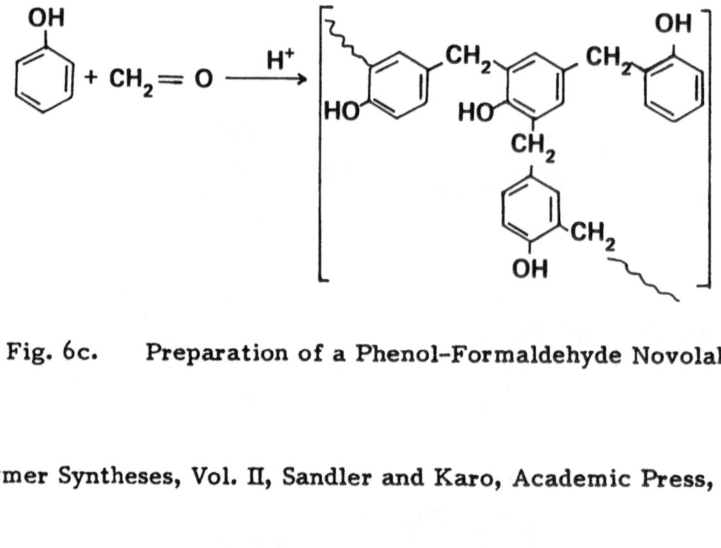

Fig. 6c.   Preparation of a Phenol–Formaldehyde Novolak

Polymer Syntheses, Vol. II, Sandler and Karo, Academic Press, 1977.

Epoxies. These materials polymerize and cross-link through a number of reaction mechanisms. For wood gluing the amine induced hardening, as shown in Fig. 7, is preferred because it can be made to proceed at room temperature and does not result in acidic glue lines.

In general, the epoxies have not found good acceptance in the wood gluing industry. There are several problem areas. Due to the mass/exo-therm effect, epoxies tend to have very short pot-lives in mass while

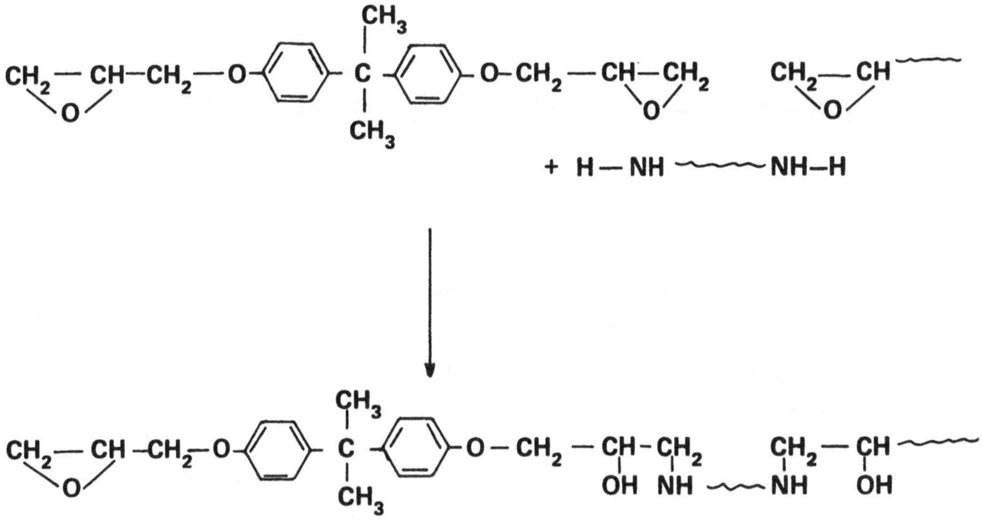

Fig. 7.   Curing of Epoxies with Amine Hardener.

needing very long cure times in thin glue lines. Equipment washability is a problem due to the need for organic solvents. Finally, many of the amine hardeners have caused dermatitis of employees.

Isocyanates/Polyurethanes. Isocyanates have been known for many years as being highly reactive materials. The isocyanate group can react (polymerize) 1) with itself, 2) with active hydroxyl groups of polyols (or cellulose), 3) with water or 4) with amines (Fig. 8). Therefore, isocyanates by themselves have been useful as binders for composites.

For wood gluing, two isocyanates (Fig. 9) have been used. m-Tolylene diisocyanate was used initially. Due to the high vapor pressure and toxicity its use remained very limited. More recently MDI ("methylene diisocyanate", pp'-diisocyanodiphenyl-methane, has found commercial use as a wood adhesive, especially for composites.

When materials such as the epoxies and isocyanates are used as sole wood adhesives, dependence on the benzene ring availability (oil) is not much changed, although these binders have been claimed to perform well at levels lower than those of phenolics or aminoplasts, resulting in some reduction of benzene consumption. Their larger benefit may ultimately come from their capability to cross-link linear or other non-cyclic polymers. In this way the possibility exists to replace well over one-half of an adhesive system by chemicals not derived from benzene. Binders of this type are now being offered on the market. I assume we will see much activity in this research area in the near future. One should question if a

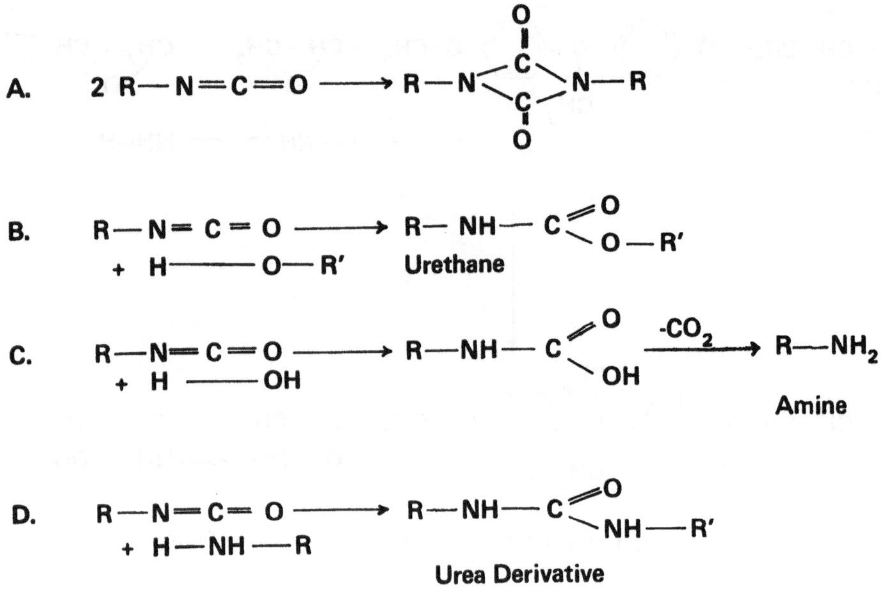

A.   2 R—N=C=O ⟶ R—N⟨C(=O)⟩N—R

B.   R—N=C=O ⟶ R—NH—C(=O)—O—R'
     + H——O—R'   Urethane

C.   R—N=C=O ⟶ R—NH—C(=O)OH ⟶$^{-CO_2}$ R—NH$_2$
     + H——OH   Amine

D.   R—N=C=O ⟶ R—NH—C(=O)NH—R'
     + H—NH—R
               Urea Derivative

Fig. 8.   Reaction of Isocyanates.

Polyurethanes, Vol. XVI, Pt. 1, Saunders and Frisch, Interscience Publishers.

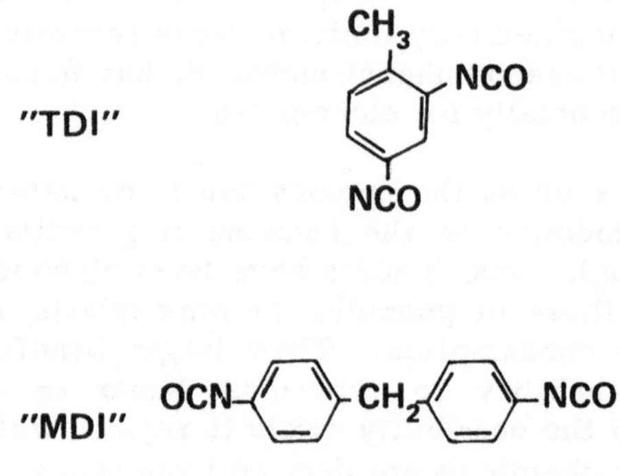

"TDI"

"MDI"   OCN—⟨ ⟩—CH$_2$—⟨ ⟩—NCO

Fig. 9.   Isocyanates Used in Wood Gluing.

suitable polyol can be prepared from a carbohydrate substrate, such as cellulose or starch.

Polyvinyl Acetate (PVA) Types. For nonstructural applications this group has enjoyed much popularity for a number of years. Although the bond quality of these adhesives is good, bond line durability under moist conditions has been marginal to poor and creep remains a major problem, even for the so-called cross-linking types.

The cross-linked PVAs were initially very successful to gain full (and excellent!) approval based on existing well established test methods (such as PS-56-73). They showed considerably better overall performance than any of the phenol-resorcinol-formaldehyde (PRF) resins. It was not until the PVAs were tested for creep that their use for structural applications became questionable. I believe some day we will be able to cross-link this thermoplastic molecule to the point where it will be able to withstand continuous loading at normal structural exposure conditions.

In some ways the polyvinyl acetate and isocyanate based compositions can be viewed as having similarity. At present very little, if any, is being used for structural bonding, but there is a good chance that further work on the chemistry of these systems will move them toward (and ultimately into) the structural area. Both technologies offer the reward of replacement of a fair portion of the benzene ring by other, more accessible raw materials.

WHERE DO WE GO

Addressing the question of the direction of future research for wood binders causes discomfort and concern.

Durability

Up to now, the basis for the durability of our wood binders and adhesives has been the benzene ring, whether in the form of phenol, resorcinol, or similar structures. The benzene ring has been the key to long-term durability, and as we move from the phenolic type over the less directly bonded aromatic networks such as epoxies and polyurethanes to the nonaromatics, we seem to lose durability.

Due to the abundant and low cost raw material supply for durable phenolic types during the past half century, we have enjoyed the luxury of being over-conservative in some applications. This luxury may slowly disappear. We must learn to target our bond durability needs more precisely and better tailor our bond durability, to offer a greater variety of bond compositions (and durabilities). At the same time, we must better guarantee our bonds. The ultimate target is a cellulose/lignin/polymer based structural product for which a minimum performance level (strength and durability) can be guaranteed to the user.

Chemistry

For our chemists, the target will be to reduce (or eliminate?) the dependence on the benzene ring while maintaining bond durability commensurate with product specifications and end use requirements. I suggest the following approaches for consideration:

1) Develop melamine chemistry to achieve low cost, low (room?) temperature cure, and to compete with present PRF adhesives overall.

2) Use lignin as a partial replacement for phenolic oligomers.

3) Learn to (catalytically) pyrolize lignin to form mononuclear phenolic species suitable as raw material.

4) Learn to attach isocyanate groups to the lignin molecule or its fragments.

5) Learn to use starch as the major raw material (in combination with phenolics). Carbohydrates have been used as adhesives for millenia. They are good adhesives or binders, but their durability (moisture resistance) is poor. Can we learn to modify (insolubilize) cellulose analogous to the cellulose-chitosan-chitin series? Or use an equivalent approach with starch?

6) Explore the use of bark catechins as replacement for phenolic species.

7) Develop PVAs and isocyanates to be fully structural.

8) Improve the durability of urea-formaldehyde (UF) resins. The Europeans have been and are using UF adhesives in 80% of their structural laminated beams.

9) Prepare hotmelt adhesive type isocyanates, forming a rapid bond as a hotmelt adhesive but later on (slowly) "curing out" due to "moisture cure"?

10) Learn to carry out in-situ conversion of carbohydrates to furan derivatives, the latter being the raw material for polymer-adhesive formation.

Application Methods

In comparison to the effort spent on the development of the chemistry of wood adhesives, the development of mechanical devices for a metered, uniform, well controlled application (quantity and placement) has been much neglected.

The general lack of cooperation between wood technologists, chemists and engineers has done much to keep us in the medieval era. If progress is to be made, these barriers must be removed. There is much to be gained by developing new metering, mixing and application methods for multi-component systems. These methods must achieve a well-metered, well-mixed composition combined with a one-trip (no recycle!) application. We then can build energy into the chemical system to cause rapid room temperature cure, eliminating the need for external energy in the form of steam, hot air, radiant or radio frequency pre- or post-heat, most of which are not very energy efficient.

There are many potential advantages to continuously metered, mixed and applied multi-component systems. The chemistry is either on hand or near, but the mechanical capability is missing. The solution of these mechanical problems may not be easy, but the rewards will be high.

IN SUMMARY

We must start a search for chemical materials, structures if you will, that will be available for the next few decades, renewable within a reasonable time frame, and resistant to breakdown by sunlight, weather, moisture, and elevated temperatures.

We then must take our wits and learn to channel the chemistry to achieve the final solidification step where we want it to happen - in the glue line.

Development of a chemical composition (material) will not be sufficient. We need to develop total systems, to combine chemistry and engineering.

Energy can, at least partially, be replaced by chemical reactivity. As an approach I suggest the development of multi-component systems combined with a no-recycle application.

Ideally, the test method correlation will dictate the type of adhesive, the "system" will apply the multi-component composition, and the formation of bond will be without further application of energy and at ambient temperature.

THE FORESTS AS A SOURCE OF NATURAL ADHESIVES

P.R. Steiner

Forintek Canada Corp.
Western Forest Products Laboratory
Vancouver, British Columbia   V6T 1X2

Natural adhesives are appealing to our present industrial
society because they conjure up thoughts of products derived from
completely renewable resources, thus offering solutions to the price
vulnerability and diminishing availability of the petroleum resource.
Sometimes, however, our thought processes are inhibited by the rapid
scientific and industrial changes which have occurred over the last
few centuries. Going back in history, we soon realize that natural
materials have been used as adhesives since at least the time of the
Pharaohs. Only in the last hundred years have synthetic-based
materials made a significant impact. In fact, it should be remembered that these so-called synthetic materials, which are produced
from petroleum or coal feedstocks, in themselves originate from
plants and animals. Under the influence of pressure, time and an
anaerobic environment, they decomposed to these hydrocarbon sources.
This process required tens of thousands of years. Our current energy
problems arise because this resource is being depleted in the relative short time period of a few hundred years. By using plant
products, the growing cycle can be matched to the time of man's
utilization. If planned carefully, these products then become
renewable (in our lifetimes) chemical and energy resources.

In contrast to refined synthetic hydrocarbons, natural products
are usually solid materials containing carbon, hydrogen, nitrogen
and/or oxygen. They consist of a multitude of molecular species
varying from monomeric to highly polymerized states. The characteristic physical and chemical properties of these materials render
them useful as adhesives.

Typically, natural adhesives are carbohydrate or protein based
and derived from either animal or vegetable sources. Much has been

written regarding their application and bonding properties (1,2).
Starches, other cellulosics, animal glues, casein and soybean
adhesives are part of this category which provides medium to low
water resistant properties when used to bond lignocellulose composites. Because a number of these materials are also food sources,
competition for these resources will intensify with population
growth.

The forest, in contrast, offers a presently under utilized source
(except possibly the latex for rubber) of raw materials for adhesives. Methods are presently available to efficiently harvest tree
boles and, with minor modifications, the other tree components also
can be recovered.

The photosynthetic transformation of carbon dioxide and water
within tree foliage produces carbohydrates, lignin and extractives.
The chloroplasts of leaves act like a chemical factory and produce
precursors which are transferred via the cambium to sites where
enzymatic or catalytic polymerization convert them to particular
polymeric compounds.

The following discussion will be focused on three major sources of natural adhesives from trees: lignin, bark and foilage.
The approximate amounts of each of these components in hardwood
and softwood trees are given in Table 1. Chemical and physical
properties that influence the present and future utilization of
these three components as adhesives will be emphasized.

Table 1. Ranges for Component Proportions in
Softwoods and Hardwoods

Component	Amounts (% dry weight of bole)	
	Softwood	Hardwood
Lignin	22-32	16-24
Bark	10-23	10-17
Foliage	10-25	5-15

1. LIGNIN

In most land plants, lignin imparts stem rigidity providing
vertical support. Lignification occurs in the cell walls, primarily
in cells having functions of water and nutrient transport or of
mechanical support. Lignification controls diffusion across the cell
walls. Also it helps wood to resist attack by micro-organisms. To

impart this important rigidity function requires an efficient composite action between lignin, cellulose and hemicellulose.

The wood pulping process breaks down this lignin-cellulose interaction and frees valuable cellulose fiber. Historically, the lignin fraction has been viewed as a nuisance, as an energy source (3) or, in recent years, as a potential feedstock material for producing aromatic compounds (4). Since nature evolved an effective adhesive system for wood, one could modestly try to learn from this process. Unfortunately, the complexity of the natural lignin structure and the limitations of our analytical methods prevent us at present from efficiently utilizing this resource. To appreciate the potential of lignin as an adhesive, we must first attempt to understand the reactivity and structure of the material itself.

Chemical Reactivity

Chemical pulping breaks the lignin macromolecule into smaller units through hydrolysis of the ether linkages connecting different lignin subgroups or present in lignin-carbohydrate bonds (Figure 1). In acid sulfite pulping, sulfonation and hydrolysis together result in lignin solubilization into the pulping liquors. The lignin is recovered as a sulfonate. Little methoxyl group loss takes place, and the sulfonate groups, which predominate in the benzylic position, impart a strong hydrophilic character to the lignin product.

In kraft pulping, the combination of strong base and sulfides cleaves some ether linkages and increases loss of methoxyl groups. Lignin molecular fragmentation is more severe than in sulfite pulping with the formation of phenolates resulting in lignin solubilization. In both pulping processes, oxidation and condensation of reactive groups can occur.

A lignin product of the future may be available from the organosol pulping process. The milder reaction conditions should reduce condensation reactions during pulping. However, the resulting product will be primarily alkyl ether derivatives or oxidized products with few free hydroxyl groups. This will retard further polymerization of these materials into cross-linked adhesive products.

It is well recognized that two primary substructures make up the lignin macromolecule, the guaiacyl propanol and syringyl propanol units. In softwood tree species only the former exists, while in hardwood species both nuclei are present. The relative reactivity of positions on these propylaryl units have been extensively studied (5). For instance, with formaldehyde, initial hydroxymethylolation takes place at position 5 (Leder-Manasse reaction) with further possible hydroxy-methylation of positions on the side chains adjacent to carbonyl groups (Tollens reaction) or at a conjugated double bond.

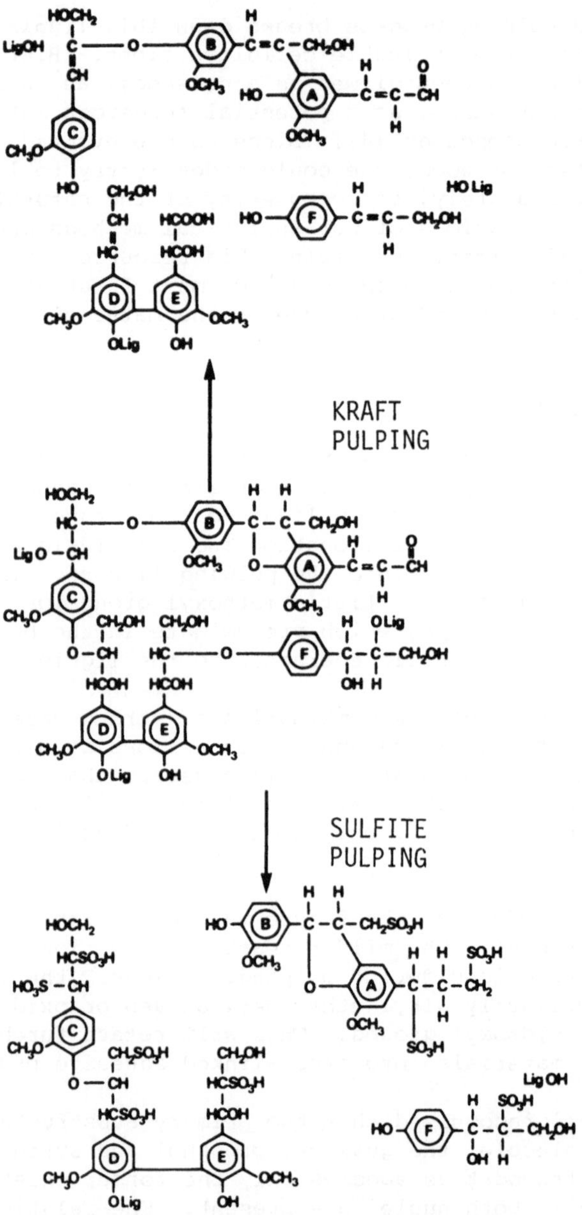

Figure 1. Idealized structure and product from the kraft and sulfite
pulping reaction of softwood lignin.

Guaiacyl propanol   Syringyl propanol

Under acid conditions, positions 6 and 2 (meta to the phenolic hydroxyl) appear most reactive, especially if demethoxylation has taken place. In lignin with a large proportion of syringyl groups, reactivity is further restricted by the methoxy group at the 5 position. Hence softwood and hardwood pulping liquors would show different response to reaction with formaldehyde. Studies show about 40 methylol groups (75% in ring, 25% in chain) are introduced into softwood kraft lignin per 100 arylpropane units (6).

Derivatization reactions involving phenol (7), isocyanate (7) and epoxides (8) have also been studied.

Physical Properties

Since delignification is influenced in part by chemical accessibility, stability of lignin carbohydrate bonds, wood variability and secondary condensation reactions, the resulting soluble lignin product is structurally complex. Table 2 shows the molecular weight ranges for some typical lignins. The kraft process produces a lower molecular-weight (m.w.) lignin than the sulfite pulping process. Such factors as molecular size, molecular shape and polydispersity influence the rheological and solubility properties of lignin which, in turn, can sterically hinder reactions. Lignin units generally have a three-dimensional structure with somewhat spherical shapes, although some asymmetric,rod-shaped units are present in high m.w. components (9,10). These shapes and the polydispersion in size suggest very strong association through secondary bonds.

Lignin is a thermoplastic material whose softening point is influenced through plasticization with water (10). Chemical processes, such as oxidation, condensation, hydrolysis, etc., which alter

Table 2. Molecular Weight Ranges for Some Typical Lignins

Tree	Pulping Type	Molecular Weight ($\overline{M}w$)	$\overline{M}w/\overline{M}n$
Spruce	Sulfite	5000-130,000	6.4
Spruce	Kraft	1800-5000	>3.0
Spruce	Dioxane and HCl	4300-85,000	3.1
Hemlock	Sulfite	450-58,000	6.4
Pine	Kraft	3500	2.2
Hardwood	Kraft	2900	2.8

functional groups during delignification, affect thermal properties of the lignin product. The ability of lignin to flow and penetrate into the wood at a specific rate in relation to temperature, time and pressure will influence its utility as an adhesive.

Utilization as an Adhesive

An early application of lignin's adhesive properties (11) was the pressing of decayed wood to give a product of high strength and some water resistance. Here, fungal attack on the cellulose portion of wood had physically and chemically restructured the lignin into an active adhesive. Similar effects have been reported with wood-hydrolysis treatments (11).

Much research effort has been applied to lignosulfonates as adhesives because of their availability and low cost (12-17). Their main drawback has been hydroscopicity resulting from the presence of sulfonic acid groups. Lignosulfonates were used commercially in the early 1960s to bond wood chips, but long pressing times followed by autoclaving was necessary to hydrolyze some sulfonic acid groups and produce waterproof boards (12). Some improvement in curing time was achieved by addition of small amounts of acid-curing phenol-formalde-hyde (PF) resin, but high pressing temperatures were still required. Shen (13,14) greatly enhanced cure by acidification of the lignosul-fonate salt with sulfuric acid to give a liquid or spray-dried ligno-sulfonic acid product. Similar effects (15) could be achieved by electrodialysis treatment. Although good bond quality was achieved in waferboard made with acidified lignosulfonates, concerns were ex-pressed about panel acidity and its effect upon long-term wood degra-dation, in addition to metal corrosion problems. With reduced acid content, bond stability is improved but with a compromising increase in cure time (14). More recent work (16) with ammonium lignosulfo-nates suggested that the reducing-sugar level in the lignosulfonate

system influenced bonding properties. The pressing conditions of
high temperature and long time likely allowed conversion of some of
these sugars to furfural derivatives which, on polymerizations,
yielded a highly thermal- and hydrolysis-stable polymer.

Another approach in utilizing lignosulfonates is to separate
the high molecular weight species by ultrafiltration and use these
complex three-dimensional polymers as a co-reactant with low molecular weight PF resin (17). The resulting resin system contains about
50% lignosulfonate. Cure time is significantly improved because a
large portion of the three-dimensional network is achieved prior to
curing.

Numerous applications of kraft lignins as adhesives have also
been reported (6,17,18). On acidification, a precipitated thermoplastic product results that can bind lignocellulose material and
provide dry strength. However, the presence of inorganic ions reduces the water resistance of the bond. Purification of the lignin
component helps remove these ions, but chemical modification through
addition or condensation reactions are also needed to significantly
improve adhesive properties (17). Even with many of these modifications, the complex structural character of lignin and its low reactivity have proven a drawback in producing a highly crosslinked,
adhesive product (18).

Both Marton et al. (6) and Dolenko and Clarke (18) have co-
reacted formaldehyde and lignin. The resulting adhesive systems
showed some improved properties, but still required addition of 30%
or more of low-molecular-weight PF resin to achieve durable bonds in
waferboard. Presumably, the smaller size and the greater reactivity
of the methylol phenols provide the appropriate crosslinking agent
for the lignin.

Overall, most attempts to use the lignin macromolecule only to
produce self-condensing durable adhesives have proven unsuccessful.
At best, about 50 to 70% of a PF resin can be replaced by lignin.
This is not surprising since the lignin product is highly variable,
having been exposed to extremes of temperature, pressure and pH
during the pulping process. This moderates reactivity and increases
structural complexity to the point of limiting the use of lignin as
an adhesive.

By breaking down the lignin macromolecule by oxidative alkaline
hydrolysis, hydrogenolysis and demethoxylation, structural complexity
and variability are reduced and reactivity may be increased. For
example, hydrogenolysis and demethoxylation of lignin can lead to
4-methyl-catechol, which reacts well with formaldehyde to produce
resins with properties similar to phenolic resins (19). Demethoxylation of lignin without hydrolysis should also enhance reactivity
with formaldehyde. For commercial applications, it is important

that any process produce simplified products in high yields. A promising recent process (20), involving the reaction of lignin in the presence of boron triflouride, has reported yields of catechol of 0.2 to 0.42 moles/OCH_3 of lignin by a combination phenolysis, hydrolysis and demethoxylation reaction. The resulting catechol provides a useful starting material for phenolic-type resins.

2. BARK

Bark, being readily available, has been used for centuries for clothing, rope, paper, cork, roofing and as a source of medicinal chemicals. Presently, large quantities of bark are available at the mill sites where it is utilized as an energy source, in landfill, or for horticultural purposes. On average, barks consist of 15 to 30% polyphenols, 20 to 30% lignin, 30 to 45% carbohydrate material and 1 to 3% fats and waxes. These values are highly dependent on species, growing site, age and position in trees. The older outer bark offers protection for trees. The younger inner bark functions as a growth interface where nutrient transfer occurs.

Because almost two-thirds of bark is composed of aromatic components, the natural availability of such substantial, low-cost polyphenolics has long interested adhesive chemists. Over the past 40 years, substantial research into the use of bark extract materials as adhesives has been undertaken in Australia, New Zealand, South Africa and North America. Some limited commercial successes have been achieved in Australia and South Africa with wattle and radiata pine species.

Chemical Reactivity and Structure

The major class of polyphenols in bark are the condensed tannins that can range from 3 to 30% of the bark. Tannin material extracted in aqueous solution often contains closely associated carbohydrate components. The basic structure of condensed tannin is C_6- C_3- C_6 flavonoid unit;

usually linked through C-4, C-6 or C-8 in its polymeric form. The
functionality of the A and B ring govern the rate of electrophilic
reaction of the particular flavonoid. The A ring can be a substituted resorcinol, X = H, as found in wattle or quebracho species, or
can be a substituted phloroglucinol, X = OH, as found in radiata
pine, southern pine and western hemlock. The B ring is usually
either a pyrogallol (Y = OH) or catechol (Y = H) group with the
latter predominating.

This difference in A-ring functionality is a significant factor
governing the working and curing properties of tannins when used as
adhesives. The relative nucleophilic reactivity, phloroglucinol >
resorcinol > pyrogallol >> catechol, is consistent with the
greater formaldehyde reactivity of pine bark tannins compared with
wattle tannins. This is supported by gelation-time studies (21) at
pH 6 where the relative rates of gelation in the presence of formaldehyde was 1: 4: 20, respectively, for wattle, mangrove and radiata
pine extracts.

Studies of formaldehyde reactivity (22) with catechin (X = OH)
and polyphenols (23), and bromine reactivity (24) with tetramethyl
catechin have shown reactions occur first at the C-8 position and
then at the C-6 position of the A ring. The B ring positions are
relatively unreactive until high alkaline conditions are achieved,
whereupon the C-6' position becomes active. Such high alkalinity,
however, increases A-ring reactivity to such an extent that the
reaction becomes uncontrollable. For resorcinol-type flavonoids
(X = H) reactivity follows the order C-6 > C-8 >> C-6'. While increasing alkalinity enhances tannin solubility and reactivity of the
B ring, such treatment can produce rearrangements to less reactive
catechinic acid structures (25) and/or oxidation to quinones.

The pyran ring is labile in mild acid or bisulfite and this
leads to electrophilic sites which may react with nucleophiles, such
as phenol or resorcinol, if present (24). Such reactions can improve
solubility, reduce viscosity and offer potentially greater reactivity
at the new reaction site.

With the phloroglucinol A ring (X = OH), it has been suggested
that the difference in reactivity between it and the catechol B
(Y = H) ring makes effective crosslinking of tannin with formaldehyde
in solution difficult (26). Hemingway (27) found that catechin condensation reactions with hydroxybenzyl alcohols are much slower than
those with formaldehyde, even at comparatively higher temperatures.
These results suggest the polymethylol phenol modification route may
provide more controllable reaction conditions for resin formation
while increasing the presence of reactive sites.

Physical Properties

Bark behaves quite differently from wood because of its unique
anatomy, large aromatic composition and low fiber content. The
presence of living and dead tissue leads to extremely variable water
contents and specific gravity ranges among and within species. Notably, the compression strength of bark along the grain is only one-
third to one-sixth the strength of wood (28). Chow and Pickles (29),
in a thermal-softening study of Douglas fir and red alder, showed dry
ground bark initially softens at 180°C, with significant additional
softening at 280°C where water of condensation was released. The
initial softening is influenced by water and can be attributed to
plasticization through reduction of interfiber hydrogen bonding. At
higher temperatures, polymerization of extractives and lignin or
degradative depolymerization occurs.

The inherent variability of bark species is reflected in the
composition and yields of bark extracts. In southern pines, flavonoid composition and yield change abruptly from phloem to outer bark
and within the outer bark vary with tissue age (26).

Tannin extract solutions are colloidal and polydispersed. Dramatic variations in particle size distributions occur with
changes in concentration, temperature, pH and added salts, with
larger particle sizes tending to be hydrophobic. The strong molecular association present in tannin solutions has resulted in some
ambiguities in molecular weight values. Table 3 shows molecular
weight ranges of several bark extracts. In some cases, the presence
of high molecular weight gums closely complexed to the tannin possibly cause abnormally high values.

Table 3. Molecular Weight Ranges of Bark Extracts

Material	M.W. Range
Wattle	$\overline{M}_n$ 1250 – 3000
Quebracho	$\overline{M}_n$ 1800
Radiata pine	$\overline{M}_n$ 8400
	$\overline{M}_w$ 10,000 – 30,000
Loblolly and	$\overline{M}_w$ 2500 – 9000
shortleaf pine	$\overline{M}_n$ 1000 – 3500
	240,000 – 500,000

Haslam (30) has suggested that bark flavonoid oligomers are in-
flexible chains formed in a helical structure imposed by the conformational restraints of the inter-flavonoid bond. The B ring projects
about the central core, comprising the A and pyran rings, thus re-
stricting some reactive sites in the structure. This immobility of
tannin molecules limits the number of potential cross-links by re-
stricting alignment of suitable reactive positions for bond formation. NMR studies (30,31) show that increased conformational mobility can be achieved by use of appropriate hydrogen-bond-breaking
solvents (i.e. nitrobenzene) or by heating to about 170°C.

These conformational restrictions may also influence the ability
of tannin to penetrate cell walls of wood for efficient adhesive
bonding.

Utilization as an Adhesive

At pressing temperatures below 180°C, bark particles can be
formed into a board with some water resistance via the plasticization
mechanism, without the addition of synthetic resin (32). Chow (33)
showed that pressing temperatures of 200°C or more produced a higher
quality exterior-type bark board due to the polymerization of phenolic extractives and lignin in the bark. Strength, dimensional stability and moisture absorption of these boards were similar to or
better than bark boards made with 4 or 7% urea formaldehyde (UF) or
PF resin at pressing schedules similar to commercial particleboard
production.

Bark powders have been utilized as the hydroxy group source in
the formation of epoxy or polyurethane systems (34) or as partial
replacement for UF or PF resins in adhesive mixes (35,36). In these
latter cases, viscosity control was a limiting factor.

Interest in aqueous bark-tannin extracts as adhesives was activated by Dalton's work (37) in Australia. Wattle extracts reacted
with formaldehyde produced water-resistant wood bonds after hot
pressing. Although these results were promising, boiling-water tests
indicated poor wood failure and viscosity control was also a problem.
Further developmental work by Plomley and co-workers (21) has shown
these adhesives can be used in the bonding of particleboard and plywood, with the latter requiring fortification by PF or phenol-res-
orcinol formaldehyde (PRF) to achieve exterior durability. Tannin
materials from the barks of quebracho, mimosa, mangrove and radiata
pine have also been used to bond lignocellulose materials with mixed
success.

Recent findings suggest that tannin extract purity has a significant influence on bond strength and water resistance. As little as
20% carbohydrate material in the extract reduced bond strength almost

by half (38). For this reason, reduction of this carbohydrate component by fractionation, precipitation or use of organic extraction
solvents has received more attention (26,39).

Compared with wattle tannins, bark extracts from spruce, hemlock
and southern pine species have the disadvantage of lower yields of
bark polyphenols and a reactivity with formaldehyde that is too rapid
to offer good adhesive working properties. This latter phenomenon
relates to the before-mentioned difference in flavonoid A-ring structure. Pot-life has been extended by using paraformaldehyde or hexa-
methylenetetramine instead of formaldehyde solution. Steiner and
Chow (40) showed by thermal analysis and plywood-bond tests that
higher temperature pressing improves western hemlock tannin extract-
formaldehyde bonds by providing sufficient energy to extend cross-
linking. Presumably this involves enhanced reactivity of the polyflavonoid B ring.

Extensive work by Roux (24), Plomley (21), Hillis (41) and
Hemingway (26) has shown that, through increased understanding of
bark chemistry and reactivity of model compounds, improvements were
made in the handling and polymerization of bark tannins. However,
advantageous situations (such as South African wattle tannin) where
forestation, growth climate, harvesting, extraction methods, and concentration techniques are highly standardized, appear necessary to
achieve bark-tannin adhesives of commercial quantity and uniform
quality. Even then modifications are necessary to increase the
degree of reactivity and reduce glueline brittleness.

Modifications have involved grafting of phenol monomers onto the
polyflavonoid and then co-reacting with formaldehyde *in situ* to produce a tannin-resol system (24) or by fortifying the tannin extract
with a PF or UF resin (36). Both approaches produce improved adhesive properties with reduced press times. The success of these modifications relates to the increased reactive sites for formaldehyde
and the ability of these low-molecular-weight additives to bridge
sterically hindered reactive sites of the larger tannin molecules.
It is still uncertain whether the fortifying resin co-reacts with the
polyflavonoid or only forms an intertwining cured matrix with the
bark polyphenols. Hemingway concluded (27) that the effective use of
PF fortifiers requires: a high methylol content PF of low molecular
weight, solution pH less than 8, methylol phenols to be partially
condensed with tannin prior to pressing, and in particular no free
formaldehyde be present during the early stages of adhesive formation.

Although advances in tannin-extract utilization have been promising during the last 25 years, product variability between trees and
species, with their inherent reactivity and steric influences, are
factors which will always be present. For these reasons, conventional pressing techniques will likely require a fortified-tannin adhe-

sive to achieve a high quality, durable bond for the foreseeable
future. However, additional knowledge of structure and its effects
on reactivity, together with higher pressing temperatures and more
effective catalysts, will afford a tannin-only adhesive for bonding
lignocellulose products of intermediate durability.

3.   FOLIAGE

     Foliage, being the site of photosynthesis and chemical transfor-
mation, supplies the raw material for tree stem growth. The requirement of mobility, to assist transport to the stem, requires that a
large portion of the phenolic, carbohydrate and protein substance in
foliage be of a simpler and smaller molecular structure compared with
similar substances in wood. Foliage also is capable of regulating
and maintaining water content in living plants.

     The foliage component in a tree can be a substantial portion of
the unbarked bole weight, depending on maturity and tree species. In
softwoods, foliage comprises about 10% of mature trees and up to 25%
of young trees. Respective figures for hardwoods are 5% and 15%,
with seasonal variation being much greater (42). Recent whole-tree
transport studies indicate that 2.8% of pine and 8.5% of spruce foliage (as a percentage of total green-tree weight) reaches the mill
site (43). Approximately one-half the foliage and branch components
are broken off during felling and skidding of trees.

     Increasing interest in foliage utilization has been evident in
recent years (42-45), with applications concentrated in the field of
animal feed ("muka") and chemicals (essential oils, vitamins, chloro-
phyll). The literature on adhesive-related applications is more
limited. A still unresolved concern is the effect a systematic removal of foliage from the forest soil will have on forest ecology.

Chemical Reactivity and Structure

     The well-known dependence of foliage growth on age, site, season
and species results in a large variability in composition values for
this material. Some general composition ranges for softwoods and
hardwoods are compiled in Table 4. Near-identical samples have resulted in differences in reported values, depending upon the analysis
method. The main differences between softwoods and hardwoods appear
to be in the protein and fat content. Also, most hardwood species
drop all their leaves annually. There is also a significant difference between hardwood and softwood foliage in their lignin and phenolic acid structures (50). Foliage from the softwoods, like the wood,
contains only the guaiacyl and lacks the syringyl groups while both
are present in the hardwoods.

Table 4.   Foliage Composition

Chemical Component	Softwoods[1] (wt.%)	Hardwoods[2] (wt.%)
Crude protein	6-12	12-20
Fats	5-12	3-8
Carbohydrate	23-36	15-30
Lignin	14-20	12-15
Polyphenols	10-18	9-12
Ash	2-3	2-3

[1]Spruce, pine fir } (45-49)
[2]Birch, aspen, maple

Solubilities of foliage components in organic and aqueous solvents are greater than those of bark. Chen and Paulistsch (51),
after sequential extraction with hexane, benzene, ether, ethanol and
water, reported total foliage extraction amounts of about 45% and 48%
for pine and spruce, respectively, while corresponding bark extracts
were 31% and 38%. Recently, extractions of white spruce and lodge-
pole pine foliage with 1% NaOH solution (52) have yielded 55% to 65%
soluble material based on dry weight.

Interesting aging effects of softwood foliage have been reported
for pine needles removed from 12- to 15-year-old trees and stored
outdoors on forest soil (46). Figure 2 summarizes these data for
several components. The loss in dry weight after 105 and 350 days is
10% and 27%, respectively. The loss of fats and waxes are under-
standable because of their tendency to break down and volatilize.
The large increase in Klason lignin content reflects on enzyme processes or catalytic mineral effects involved in the decomposition
process. That the decomposition process is rapid, and favours increasing the aromatic rings content, may offer advantages in the use
of this material as a durable adhesive.

The presence of protein, polyphenol extractives and lignin components in foliage all offer potential adhesive properties, as indicated in previous discussions. Of particular interest is the likely
interaction of many of these components with formaldehyde. Although
intensive studies of foliage components have been limited and more
often directed at questions of physiological interests, many of these
components are potentially reactive because of their role as building
blocks for trees. The low degree of polymerization of many foliage
components also offers the potential for less steric hindrance during
subsequent polymerization reactions.

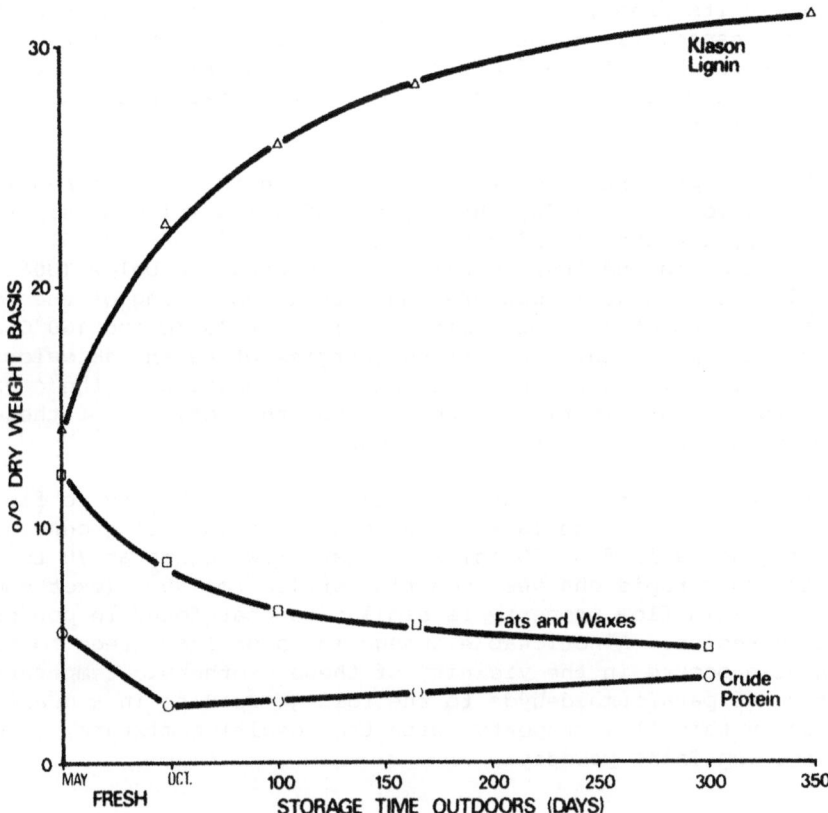

Figure 2. Changes in Pine foliage component during aging. (46).

Physical Properties

    Softwood needles have anatomical characteristics (48) of three
distinct regions: a hard outer sheath that serves as a protective
layer; a large-volume, soft-tissue zone which contains the chloroplasts; and a fibrous structural core. Their strength properties
differ from other tree components with, for example, loblolly pine
needles (53) having about half the tensile strength and an eighth of
the modulus of elasticity of wood. Foliage tends to be highly
elastic in its fresh, green state and retains some of this property
when dry, especially those tree species with substantial wax content.
This elastic character affects the ability to grind this material to
a powder. However, inclusion of some brittle twig material can enhance comminution.

    Thermal properties of foliage, as shown in the differential
scanning calorimetry (DSC) thermograms of dried white spruce foliage
(Figure 3), are characteristic of many softwood species (54). The
appearance of an endothermic peak at a temperature below 100°C represents loss of residual moisture and initial softening of the foliage
substance. Two other exothermic reactions at 130°C and 160°C also
appear with approximate respective energies of 69 and 48 calories per
gram of foliage. Although the relative intensities of these exothermic peaks depend slightly on the species, the position of these peaks
do not change with different softwoods.

    A characteristic feature of foliage is its ability to flow under
heat and pressure. Thermal-softening studies (55) of ground, dried
foliage under a load of 50 psi show that flow begins at 70 to 80°C,
with the most rapid changes occurring at 120 to 150°C (exothermic
region). This flow property is similar to that found in powdered
phenolic resins. A noticeable change in color from green to dark
brown also occurs in the vicinity of these exothermic temperatures.
Addition of paraformaldehyde to the foliage results in a drastic reduction in this flow property, with the resulting mixture, after
heating, remaining powdery.

Utilization as an Adhesive

    Foliage has been utilized as a fiber source for pulp board and
particleboard-type material (48,51,56), in both cases with synthetic
resins employed as binders. The swelling and transverse tensile
strengths of boards containing tree needles were poorer than wood-
only boards due to bonding difficulties (51). However, subsequent
formaldehyde liberation was reduced in the presence of needles.
Needles containing higher tannic acid and phlobaphene content released less formaldehyde than foliage with lower polyphenolic material. The lack of successful bonding in the needle-containing board
may be attributed to the larger particle size (1 mm) of the foliage

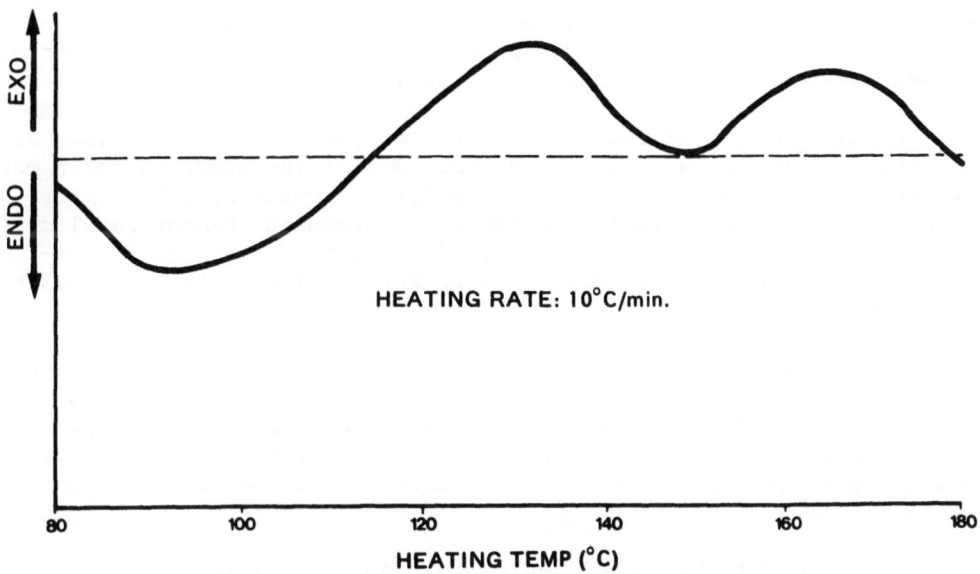

Figure 3. DSC thermogram of Spruce foliage.

material, resulting in less surface area exposure of the potential
reactive inner foliage portion.

Boards constructed from teak leaves and pressed to specific
gravity of 1.2, without additives, displayed bending strength of 181
kg/m^2 and water absorption of 21% after 24 hours of soaking (57).
Autoclaving the leaves in the presence of 10% phenol followed by
pressing resulted in similar bending strengths, but in a decrease of
water absorption to 5%.

Ground spruce and pine foliage, intermixed in a ratio of 1 to 5
with wood particles, can produce a board (58) with good dry strength
after 5 minutes press time at 150°C. Water resistance increases with
increasing press time and/or temperature. Increased foliage content
helps improve internal bond strength of boards. However, a corresponding reduction in bending strength results because of the shorter
fibers present in foliage compared with wood.

Co-mixing of powdered phenol-formaldehyde resin with foliage as
a binder for waferboard has shown about 30% of the powdered resin can
be replaced without sacrificing durability. Considering the importance of the strength factor in composite products, recent results
have shown (58) that substitution of 70 to 80% of the resin by foliage still produces adequate strength for meeting certain particleboard requirements.

The starch-like components and humectant properties of foliage
have allowed foliage powder to be used as a direct substitute for
wheat flour extender in phenol-formaldehyde and urea-formaldehyde
plywood glue mixes (54,55). Water extracts of red alder leaves also
produce peel strength properties approximating those of starch paper
adhesives (52).

Compared with bark and lignin, foliage utilization as an adhesive is, at the present time, in its infancy. Knowledge of what
gives foliage these adhesive properties is limited. Further understanding of the chemical and physical properties of this material,
in the future, will increase the efficiency of its use as an adhesive.

CONCLUSIONS

Abundance, a complimentary multi-harvesting use for fiber,
energy and chemicals, and its renewable constitution, mark the forest
as an important resource for composite products. The role of lignin,
bark and foliage as adhesives has already been partially defined by
both laboratory and industrial experience. The future use of these
forest-based adhesives may well depend upon further improving our
understanding of the chemical and physical properties of these

systems. In particular, raw material variability (both between
species and individual trees), aging effects, material separation
and purification, and the influence of co-reactants need further
study. Enhanced utilization will be attained through either improved raw-material-handling methods to retain desirable properties,
or through chemical and thermal treatment to reduce molecular size
and/or improve chemical reactivity.

Further consideration could be given to the possibility of establishing more moderate criteria for acceptable bonding in composite
products, especially with respect to their end uses. In this regard,
natural-product adhesives by themselves may provide sufficient bond
quality for some products, while other products may require fortification with more highly cross-linked synthetic resins.

REFERENCES

1. R. Houwink and G. Salomon, Adhesion and adhesives, Vol. 1,
    Elsevier Co., N.Y. (1965).
2. I. Skeist, Handbook of adhesives, 2nd ed., Van Nostrand-
    Reinhold, N.Y. (1975).
3. K.V. Sarkanen and C.H. Ludwig, ed., Lignins - occurrence, forma-
    tion, structure and reactions, Wiley-Interscience, N.Y.
    (1971).
4. W. Schweers, Utilization of lignins isolated under mild condi-
    tions from wood or wood waste for the production of useful
    chemicals and other chemical products. Presented Eighth
    World Forestry Congress, Jakarta, Indonesia (1978).
5. G.G. Allan, *In:* Lignins - occurrence, formation, structure and
    reactions; K.V. Sarkanen and C.H. Ludwig, ed., Wiley-Inter-
    science, N.Y., p511-573 (1971).
6. J. Marton, T. Marton, S. Falkehag and E. Alder, *In:* Lignin
    structure and reactions; J. Marton, ed., Advances Chem. Ser.
    No. 59, Amer. Chem. Soc., Washington, D.C., p.125-144 (1966).
7. K. Kratzl, K. Buchtela, J. Gratzl, J. Zauner and O. Ettings-
    hausen, Lignin and plastics. The reactions of lignin with
    phenol and isocyanates, Tappi 45(2):113 (1962).
8. S. Tai, J. Nakano and N. Migita, Studies on utilization of
    lignin. V. Adhesive from lignin epoxide, Nippon Mokuzai
    Gakkaishi 13:257 (1967).
9. S.I. Falkehag, Lignin in materials, Applied Polymer Symp. 28:247
    (1975).
10. D.A. Goring, *In:* Lignin - occurrence, formation, structure and
    reactions; K.V. Sarkanen and C.H. Ludwig, ed., Wiley-Inter-
    science, N.Y., p.695-768 (1971).
11. R.C. Gupta, Adhesives from wood as substitutes for synthetic
    glues, World Consultation Wood Based Panels, New Delhi,
    F.A.O. Doc. No. 33, p.7 (1975).
12. T.A. Johansen, Particleboard bonded with sulphite liquor, Proc.

Fifth Particleboard Symp., Wash. State Univ., Pullman, WA
p.11-30 (1971).

13. K.C. Shen, Modified powdered spent sulfite liquor as binder for
    exterior waferboard, Forest Products J. 24(2):38-44 (1974).

14. K.C. Shen, Spent sulphite liquor binder for exterior waferboard,
    Forest Products J. 27(5):32-36 (1977).

15. J.M. Holderby, H.S. Olsen and W.H. Wegener, Thermosetting adhe-
    sive from electrodialyzed lignosulfonates, Tappi 50(9):92-94A
    (1967).

16. K.C. Shen, Ammonium based spent sulfite liquor binder systems
    for waferboard manufacture, Proc. Thirteenth Particleboard
    Symp., Wash. State Univ., Pullman, WA. (1979).

17. K.G. Forss and A. Fuhrmann, Finnish plywood, particleboard,
    and fiberboard made with a lignin-base adhesive, Forest
    Products J. 29(7):39-43 (1979).

18. A.J. Dolenko and M.R. Clarke, Resin binders from kraft lignin,
    Forest Products J. 28(8):41-46 (1978).

19. G.E. Troughton, J.F. Manville and S. Chow, Lignin utilization
    II. Resin properties of 4-alkyl substituted catechol com-
    pounds, Forest Products J. 22(9):108-110 (1972).

20. M. Funaoka and I. Abe, The reaction of lignin under the presence
    of phenol and boron trifluoride. I. On the formation of
    catechol from MWL, dioxane lignin and kraft lignin, Japan
    Wood Research Soc. J. 24(4):256-261 (1978).

21. K.F. Plomley, Tannin-formaldehyde adhesives for wood. II.
    Wattle tannin adhesives, CSIRO Div. Forest Prod. Technol.
    Paper No. 39, Melbourne, Australia (1966).

22. W.E. Hillis and G. Urbach, The reaction of (+)-catechin with
    formaldehyde, J. Appl. Chem. 9:474-482 (1959).

23. W.E. Hillis and G. Urbach, Reaction of polyphenols with formal-
    dehyde, J. Appl. Chem. 9:665-673 (1959).

24. D.G. Roux, D. Ferreira, H.K. Hundt and E. Malan, Structure,
    stereochemistry, and reactivity of natural condensed tannins
    as basis for their extended industrial application, Appl.
    Polym. Symp. No. 28, p.335-353 (1975).

25. K.D. Sears, R.L. Casebier, H.L. Hergert, G. Stout and L.E.
    McCandlish, The structure of catechinic acid. A base re-
    arrangement product of catechin, J. Org. Chem. 39:3244-3247
    (1975).

26. R.W. Hemingway, Proc. Complete-tree util. southern pine symp.;
    C.W. McMillen, ed., New Orleans, LA., p.443-457 (1978).

27. R.W. Hemingway, G.W. McGraw and J. Karchesy, Condensation of
    ortho and para hydroxbenzyl alcohols with catechin as a model
    for use of methylolphenols as crosslinking agents in conifer
    bark polyflavonoid adhesive formulations, Symp. Chem. Pheno-
    lic Resins, Weyerhaeuser Co., Tacoma, WA. (1979).

28. R.T. Lin, Behavior of Douglas-fir bark components in compres-
    sion, Wood Sci. 6(2):106-111 (1973).

29. S. Chow and K. Pickles, Thermal softening and degradation of
    wood and bark, Wood Fiber 3(3):166-178 (1971).

30. E. Haslam, Symmetry and promiscuity in procyanidin biochemistry, Phytochemistry 16:1625-1640 (1977).

31. L.J. Porter, Structure of polymeric proanthocyanidins; elucidation of their structure and their relationship to the condensed tannins of bark, Symp. Extractives: Util. Problem or Fine Chem. Resource? ACS/CSJ Chem. Congress, Honolulu, HI (1979).

32. C.H. Burrows, Particle board from Douglas-fir bark - without additives, Ore. Forest Res. Center, Inf. Circ. 15, Corvallis, OR, 40p. (1960).

33. S. Chow, Bark boards without synthetic resins, Forest Products J. 25(11):32-37 (1975).

34. A. Pizzi, Tannin-based polyurethane adhesives, J. Appl. Polym. Sci. 23:1889-1891 (1979).

35. H.M. Saayman and J.A. Oatley, Wood adhesives from wattle bark extract, Forest Products J. 26(12):27-33 (1976).

36. A. Pizzi, The chemistry and development of tannin-urea-formaldehyde condensates for exterior wood adhesives, J. Appl. Polym. Sci. 23:2777-2792 (1979).

37. L.K. Dalton, Tannin-formaldehyde resins as adhesives for wood, Aust. J. Appl. Sci. 1:54-70 (1950).

38. A. Pizzi and D.G. Roux, The chemistry and development of tannin-based weather- and boil-proof cold-setting and fast-setting adhesives for wood, J. Appl. Polym. Sci. 22:1945-1954 (1978).

39. K.F. Plomley, W.E. Hillis and K. Hirst, The influence of wood extractives on the glue-wood bond. I. The effect of kind and amount of commercial tannins and crude wood extracts on phenolic bonding, Holzforschung 30:14-19 (1976).

40. P.R. Steiner and S. Chow, Some factors influencing the use of western hemlock bark extracts as adhesives, Proc. IUFRO Conf. Wood Gluing, Madison, WI p.61-78 (1975).

41. W.E. Hillis, Natural polyphenols (tannins) as a basis for adhesives, Symp. Chem. Phenolic Resins, Weyerhaeuser Co., Tacoma, WA (1979).

42. J.L. Keays, Biomass of forest residuals, AIChE Symp. Series 71(146):10-21 (1975).

43. G.M. Barton, J.A. McIntosh and S. Chow, The present status of foliage utilization, AIChE Symp. Series 74(177):124-131 (1978).

44. J.L. Keays, Foliage. Part I. Practical utilization of foliage, Appl. Polym. Symp. 28:445-464 (1976).

45. G.M. Barton, Foliage. Part II. Foliage chemicals, their properties and uses, Appl. Polym. Symp. 28:465-484 (1976).

46. O. Theander, Leaf litter of some forest trees; chemical composition and microbiological activity, Tappi 61(4):69-72 (1978).

47. K.N. Law, S.N. Lo and Z. Koran, Utilization of spruce foliage. Extraction of spruce protein and chlorophyll-carotene, Wood Sci. 11(2):91-96 (1978).

48. K.N. Law and Z. Koran, Utilization of white spruce foliage. Pulp characteristics, Wood Sci. 12(2):106-112 (1979).

49. J.L. Keays and G.M. Barton, Recent advances in foliage utiliza-
     tion, Can. Forest. Serv., Western Forest Prod. Lab., Inf.
     Rep. VP-X-137, Vancouver, B.C. (1975).
50. J.B. Harborne, Phytochemical methods, Chapman and Hall, London,
     p.40 (1973).
51. T-Y Chen and M. Paulitsch, The extractives of needles, bark,
     and wood of pine and spruce and their effect on particle-
     board made thereof, Holz als Roh-und Werkstoff 32:397-401
     (1974).
52. P.R. Steiner. Unpublished results.
53. E.T. Howard, Properties of southern pine needles, Wood Sci.
     5(4):281-286 (1973).
54. S. Chow, Foliage as adhesive extender: a progress report, Proc.
     Eleventh Particleboard Symp., Wash. State Univ., Pullman,
     WA   p.89-98 (1977).
55. S. Chow and P.R. Steiner. Unpublished data.
56. E.T. Howard, Needleboards - an exploratory study, Forest Prod-
     ucts J. 24(5):50-51 (1974).
57. N.C. Jain and R.C. Gupta, A note on the complete utilisation of
     trees, Indian Forester 95:841-848 (1969).
58. S. Chow, L. Rozon and P.R. Steiner, Efficiency of coniferous
     foliage as extender for powdered phenolic resin, Proc.
     Thirteenth Particleboard Symp., Wash. State Univ., Pullman,
     WA (1979).

PLENARY SESSION II

ADHESION AND COMPOSITES

ADHESION AND ADHESIVES:INTERACTIONS AT INTERFACES

H. Schonhorn*
Bell Laboratories
Murray Hill, New Jersey 07974

INTRODUCTION

This paper describes some personal ideas concerning adhesion that have evolved over a period of years. Further, by posing several important questions, we shall highlight some current areas of interest in adhesion science:

1. Why do we surface treat polymers in order to join them adhesively?

2. What happens when a polymer is surface treated?

3. Do we have to surface treat a polymer in order to adhesively bond it, or is it that we don't know enough about how to form a surface?

In addition, one other important area that will be addressed is with respect to the permanence of an adhesively bonded structure. The rising cost of materials and of bonded structures has made this a rather important consideration within the last few years. How, for instance, can we prevent delamination of bonded structures in environments which may be hostile to either one or both of the components?

*Editors Note: Because of his indisposition just prior to the Meeting this paper is an edited transcript of the author's intended presentation.

THEORIES OF ADHESION

    Various theories of adhesion have been advanced over the last few years. Since for virtually every investigator in the field of adhesion science there is a theory of adhesion, this discussion will be confined to the most important theories, viz. those that have been given the most credence in the last few years. These theories are summarized in Fig. 1.

    1. ELECTROSTATIC:
              Adhesive - Adherend System is
        Capacitor (Surface Phenomenon)

    2. DIFFUSION:
              Adhesive and Adherend Diffuse
      into each other (Volume Phenomenon)

    3. ADSORPTION:
              Adhesive is Adsorbed by
      Adherend (Surface Phenomenon)

Fig. 1 Three Theories of Adhesion

    (i) Electrostatic Theory   This theory has gained a great deal of prominence in the Soviet Union, mainly through the efforts of Derjaguin. The electrostatic theory of adhesion considers the adhesive and the adherend as a capacitor, in other words, there is a charge separation across the interface.

    Of paramount importance in adhesion science is the question of predictability. One important feature that appears to be lacking in the electrostatic theory of adhesion is the ability to predict whether or not A will bond to B, and whether A will form a strong adhesive joint to B. Firstly let us define adhesion and adhesive joint strength as they will be applied throughout this paper. Adhesion forces and interactions are involved in the bringing together of two surfaces. For example, a solid substrate (A) and a liquid (B) to be used as an adhesive, which in time will solidify to form an AB composite. In the formation of the composite we have to consider surface interactions, surface tension of the solid, surface tension of the liquid, the viscous behaviour of the liquid and the kinetics of wetting. By adhesive joint strength, we are concerned with the forces involved in the breaking apart of a composite structure. In other words, what might the stress distribution be across that adhesive joint.

    Our concern now is not only the surface properties, as they would be in, for example, joining of materials, but with bulk properties of materials and how they might respond to an applied stress. So, the making of an adhesive joint is associated with adhesion phenomena, and the breaking of an adhesive joint is associated with joint strength. Unfortunately in the adhesion literature, the latter are implied to mean different things to different investigators because the area of adhesion science is so interdisciplinary.

(ii) <u>Diffusion Theory</u>   This theory has gained a great deal of prominence and is attributed to the work of Voyutski and coworkers. In this theory one thinks in terms of polymer molecules diffusing beyond the substrate interface to form essentially mechanical locks or grips to the surface structure. One of the difficulties with the diffusion theory of adhesion is that it becomes conceptually very difficult to think in terms of the molecules of a polymer diffusing beyond the interface of hard inpenetrable substrates such as metal oxides. Another, perhaps more constraining difficulty of the diffusion theory of adhesion is the lack of predictability. The author views the diffusion theory of adhesion as the diffusion of the adhesive to the interface, not necessarily beyond the interface. In many cases chemical reactions may take place at the interface which would have a deliterious effect on the joint strength performance.

(iii) <u>Adsorption Theory</u> This is the third and perhaps the most important theory from a very practical point of view. Largely it has received the greatest impetus from work of Zisman and his colleagues at the Naval Research Laboratory in the United States. The theory is primarily concerned with the formation of an adhesive joint. The first two theories mentioned viz. the electrostatic and the diffusion theories, are concerned with the breaking of an adhesive joint, while the wettability/adsorption theory of adhesion is concerned mainly with the making of an adhesive joint. The wettability approach is concerned with the surface energy of the solid, surface tension of the liquid, viscous behaviour of the liquid, and parameters involved in the making of an adhesive joint. There appears to be two distinct theoretical approaches associated with the making and the breaking of adhesive joints. In the remainder of this paper it will be shown how both of these approaches are viable and how one can draw from each of these theories of adhesion to form a rather simple coherent picture of what is involved in the formation of a strong adhesive joint.

Fig. 2 shows a rough surface being joined to another rough surface, and two ideally planar surfaces being joined. For example, if we had two ideally planar surfaces as one does in cleaved mica crystals, you can recover the joint strength by reforming that interface in vacuum. If contacts over the entire real surfaces of the solids are achieved, formation of a strong joint would ensue, provided that the surface region contribution, in terms of its mechanical properties, reflects the bulk strength of the material. Unfortunately, in a real situation we are dealing with rough surfaces even at a molecular scale. Typically, a piece of fire polished glass will have peak to valley surface roughness of the order of 400A. Therefore, when one joins two solid surfaces, one has only a small real area of contact and consequently requires an adhesive to provide a mechanical

IDEAL SURFACES

clean, atomically smooth, planar

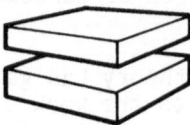

perfect (complete) contact
STRONG JOINT

REAL SURFACES

dirty, rough, non-planar

imperfect (incomplete) contact
WEAK JOINT

Fig. 2   (a)   Ideal surfaces: clean, atomically smooth,
planar;
(b)   Real Surfaces: dirty, rough, non-planar.

continuum from one phase to the other. The adhesive must conform
to the surface roughness of both substrates. By some
solidification process, whether it be similar to an epoxy which
cures by crosslinking or a thermoplastic which solidifies when
being cooled from a melt, one would form a strong adhesive joint.
So, the primary function of an adhesive is to provide a mechanical
continuum from one phase to another.

    What are the practical requirements for forming a strong
adhesive joint? Fig. 3 gives several simple guide-lines to
follow. Absence of a weak boundary layer suggests that the
surface region is mechanically similar to the bulk of the
material. The substrate should be wet by the adhesive.
Ideally the surface tension of the liquid that we use as an
adhesive should be less than the critical surface tension of
setting of the solid substrate. This is associated with the
adsorption theory of adhesion or in the formation of a strong

·   ABSENCE OF WEAK BOUNDRY LAYERS
·   FLUID ADHESIVE HAVING $\gamma_L <$ γ_c OF
SUBSTRATES
·   FORMATION OF EXTENSIVE INTERFACIAL
MOLECULAR CONTACTS
·   SETTING OF ADHESIVE

Fig. 3   Requirements for practical adhesive bonding.

Fig. 4   Schematic showing drops of six different liquids on a
         Teflon surface which dramatizes the importance of surface
         free energy in adhesives. Surface free energy of drops
         increases from left to right. Only those liquids at far
         left have a low enough surface free energy to spread on
         the low energy Teflon surface as an adhesive should.

adhesive joint. If the surface tension of the liquid is lower
than the critical surface tension of the wetting of the solid,
there will be formation of extensive interfacial contact as a
result of having the liquid adhesive conform to the surface
roughness of the solid, by displacing air from cracks and
crevices. Following the wetting process, the adhesive solidifies.
Stated simply, we would like to have the absence of a mechanically
weak surface layer, and also for the adhesive to wet the substrate
adequately. With these two notions, one should be able to predict
how to go about making a strong adhesive joint.

     How do we classify substrates with respect to their wetting so
that we can make some prediction about forming a strong adhesive
joint? In other words, how do we classify polymeric materials
with respect to their being able to be wet by a variety of liquid
adhesives. Fig. 4 illustrates the concept of wetting using a
substrate such as Teflon on which is deposited sessile drops of
various C_6 to C_{16} normal hydrocarbons. It is readily apparent

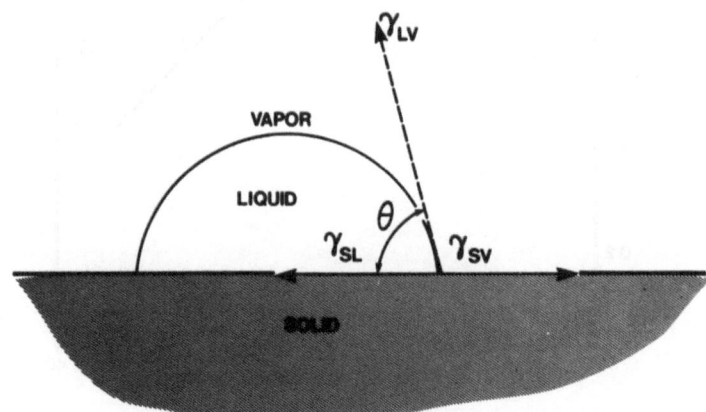

Fig. 5   Schematic of sessile drop profile and the relationship
         between the three interfacial tensions and the contact
         angle.

that the droplets exhibit different profiles. The drop profiles
increase on the surface as one proceeds to the higher surface
tension liquids, i.e. in the direction of longer chain length. If
one examines the profiles of the sessile drops on the surface,
there are three interfacial tensions to consider (Fig. 5): a
solid-vapor interfacial tension, a solid-liquid interfacial
tension which is a measure of the liquid in contact with the
substrate and a liquid-vapor interfacial tension which is
associated with the liquid in contact with its vapor. If one
draws a tangent of that drop to the surface, the enclosed angle
that the liquid makes to that surface is called the "contact
angle", θ.

A plot of the cosine of the contact angle, θ. vs the surface
tension of the wetting liquid is illustrated in Fig. 6 and
represents the pioneering work of Zisman. Extrapolation of this
line to $\cos\theta = 1$, i.e. $\theta = 0$, yields a value for the critical
surface tension of wetting, γ_c.

The meaning of the critical surface tension of wetting is that
if the liquid has a surface tension lower than the critical
surface tension of wetting, it will spread, and by spreading we
imply thermodynamic spreading i.e. giving zero contact angle. If

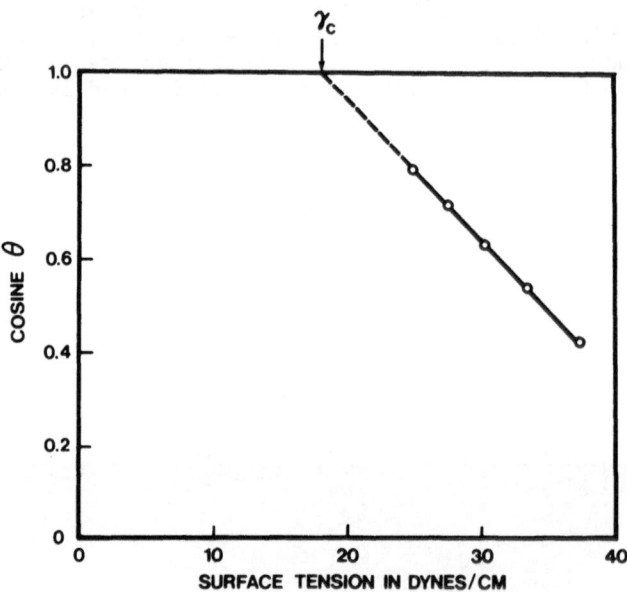

Fig. 6   A Zisman plot of the cosine of the contact angle as a
         function of the surface tension of the wetting liquid.
         The extrapolated point, ($\cos\theta=1$), is defined as the
         critical surface tension of wetting.

the surface tension of the liquid is greater than the critical
surface tension of the solid, then the angle will be finite and
the liquid may not wet and conform well to the surface roughness
of the solid. Thus, in order to wet and form a well ordered
interfacial region, the liquid must conform to the surface of the
solid. From a knowledge of the critical surface tension of the
solid, and the surface tension of the wetting liquid, one can
predict how well that liquid might wet the substrate, no matter
how rough it might be.

Since the critical surface tensions of wetting are usually
determined with simple liquids, such as water, alcohol and some
low molecular weight species, the contact angles are generally
achieved in a reasonably short period of time (seconds). However,
in real bonding situations when one is dealing with high polymer
melts, or epoxies, not only does the surface tension have to be
low, but also the viscosity. Figure 7 illustrates the wetting of
a surface with a high viscosity liquid. What becomes important is
the kinetics of wetting, i.e. how long does it take for a high
viscosity liquid to reach an equilibrium situation (Fig. 7b), and
essentially spread on the solid surface. One has to be concerned

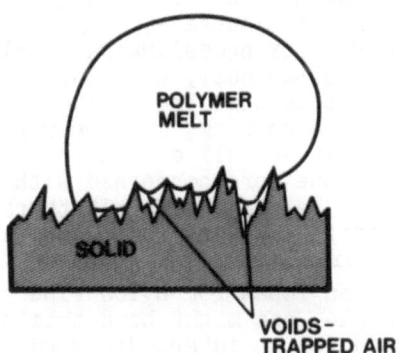

a) NOTE LOW REAL AREA OF INTERFACIAL
   CONTACT

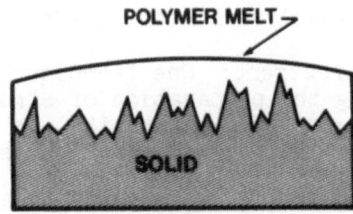

b) NOTE LACK OF VOIDS FROM TRAPPAD AIR
   IN PORES AND CREVICES

Fig. 7 (a) Poorly wetted surface; note real area of interfacial
        contact; (b) extensive intermolecular contact; note lack
        of voids from trapped air in pores and crevices.

with the ability of the polymer melt, under the conditions of high
temperature, to survive those conditions without degrading.
Polymers may degrade by crosslinking, or scission. The
thermodynamics of the absorption theory, indicate a tendency for a
liquid to wet a surface, but practical consideration may dictate
that under the conditions of the experiment, good wetting may not
be achieved.

To reiterate, the two simple guidelines to follow in preparing
strong adhesive joints are to eliminate weak boundary layers, and
to wet the substrate effectively. We surface treat polymers in
order to remove mechanically weak surface layers by increasing the
cohesive strength of the surface region. There are many polymers
that are relatively simple to wet, such as nylons, but are
difficult to join adhesively because they probably have
mechanically weak surface layers.

SURFACE MODIFICATION

In conventional surface treatments for polyolefins, i.e. glass
cleaning solution, corona discharge treatment, etc., not only is
the critical surface tension of wetting changed by introducing
polar functionalities into the surface, but they change the
constitution of the surface by crosslinking. Gelation of the
surface region occurs simultaneously with the changes in
wettability. It is suggested that the surface treatments
generally used for adhesive bonding, in some way modify the
surface region mechanically as well as change the wetting
characteristics. But if one were concerned with only wetting
characteristics, it would be very difficult to rationalize why the
joint strength for unmodified nylons is so low, since nylon has
the highest γ_c of commercially available polymers. If the proper
surface treatment had been found for nylon fibers used in the tire
cord industry, for example, one might bond them directly to the
rubber, without the need for an intermediary adhesive. It is just
that since nylon is relatively easy to wet, surface treatments are
usually avoided. Wetting would appear to be a necessary, but not
an excluseve condition for forming a strong adhesive joint. What
is required as well is to modify the surface so it can support the
large stresses that are involved in strong adhesive joints.

(i) Glow discharge treatment: One technique which can modify
polymer surfaces enabling the preparation of strong adhesive
joints, without changing γ or the wettability, is known as the
CASING technique. The CASING technique, which is an acronym for
Crosslinking by Activated Species of Inert Gases, is based on a
glow discharge treatment. For example when polyethylene is
exposed to a helium discharge we find that the joint strength
increases (Fig. 8) by a factor of 4 or 5 in a relatively short
period of time (<30 sec.), a period too short to change the
wettability significantly.

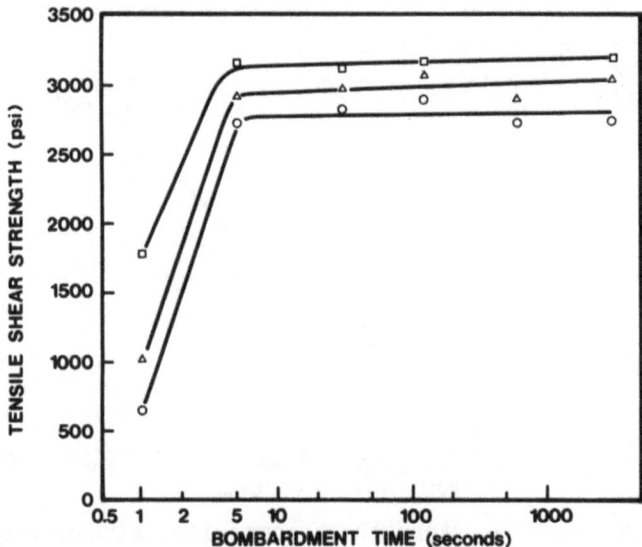

Fig. 8 The tensile shear strength of the composite aluminum-epoxy
adhesive-helium CASING polyethylene-epoxy adhesive-
aluminum is plotted as a function of the bombardment time
with activated helium at a 1 mm pressure. (O) 60°C; (△)
82°C; (□) 104°C, for three different cure temperatures of
the epoxy adhesive.

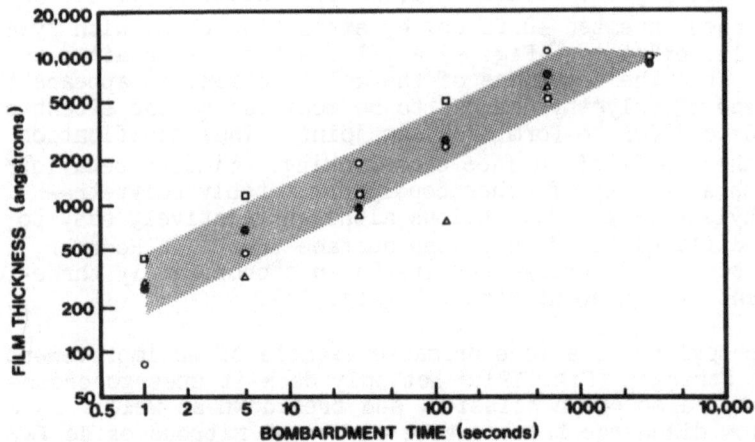

Fig. 9 The estimated film thickness as determined from the gel
fraction after Soxhlet extraction is plotted as a function
of bombardment time for a variety of activated gases: (□)
hydrogen (1 mm pressure); (●) helium (1 mm pressure); (△)
helium (0.4 mm pressure); (O) neon (1 mm pressure).

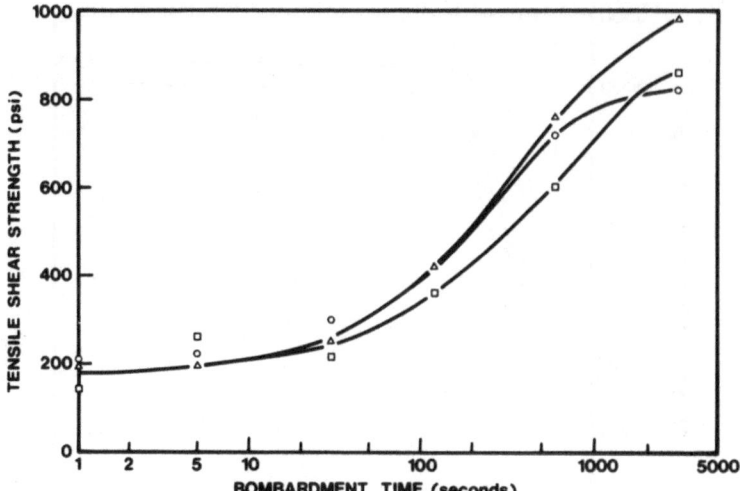

Fig. 10   Tensile shear strength of composite, aluminum-epoxy
          adhesive-polytetrafluoroethylene-epoxy adhesive-aluminum
          plotted as a function of bombardment time in activated
          neon at 1 mm pressure and high power: (O) 60°C; (△) 82°C;
          (□) 104°C, for three different cure temperatures of the
          epoxy adhesive.

    Thus the surface of the polyethylene can be modified to make a
strong adhesive joint without necessarily changing γ_c. If one
examines these treated surfaces, by extracting them, with xylene
in a Soxhlet extractor (Fig. 9) a gel fraction is obtained as a
residue. From the thickness of the gel fraction, it appears that
the surface of polyethylene has to be modified to the extent of
about 500 to 1000A to form a strong joint. Thus modification
must be the result of surface crosslinking. Similar behaviour is
noted with a variety of other copolymers notably polytetra-
fluoroethylene (Fig. 10). Nylons although relatively easy to
set, are difficult to join unless surface treated. However,
exposure to slow discharge results in an approximately three-fold
improvement in the joint strength (Fig. 11).

    Polypropylene is a more dramatic example of an improvement in
the joint strength (Fig. 12). Not only does it undergo cross-
linking, but also chain scission, and breakdown as well. In this
case a glow discharge treatment in oxygen or nitrous oxide favors
crosslinking over scission.

    (ii) Direct Gaseous treatment: The latter examples have
indicated that strong adhesive joints can be prepared without
necessarily changing γ_c however some experimental treatments which
decrease γ_c still produce strong adhesive joints. This can be

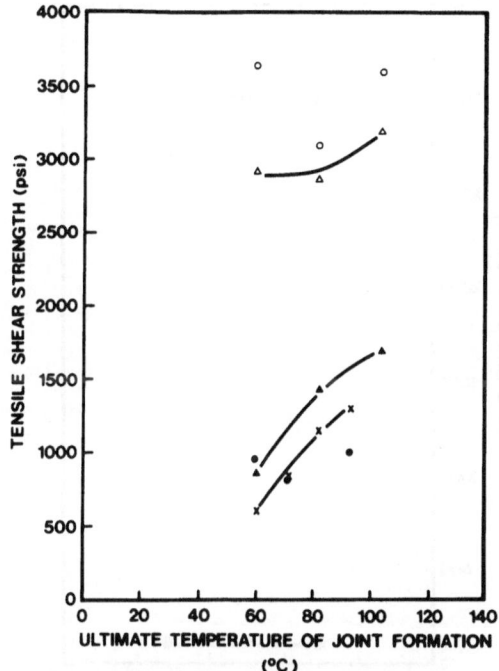

Fig. 11   Tensile shear strength of the composite, aluminum-epoxy
adhesive-nylon 6 film-epoxy adhesive-aluminum, as a
function of the ultimate temperature of joint formation:
(●) nylon 6 (Capran 77A) as received; (▲) nylon 6 (Capran
77C) as received; (O) subjected to CASING in helium for 30
min. (Capran 77A); (△) subjected to CASING in helium for
30 min. (Capran 77C).

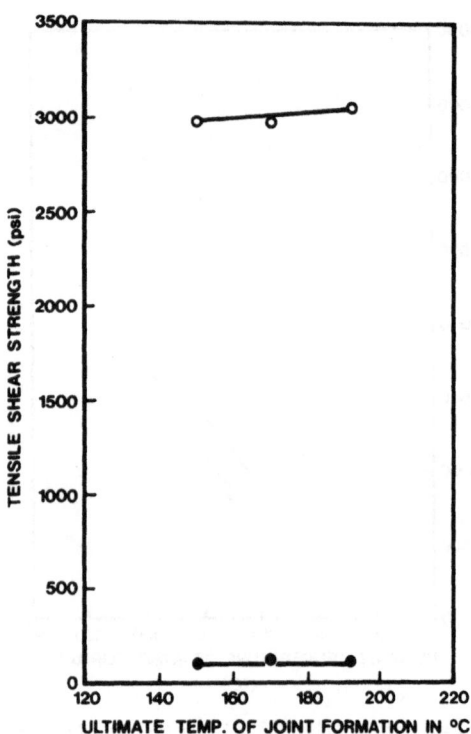

Fig. 12 Tensile shear strength of composite, aluminum-epoxy
        adhesive-polypropylene-epoxy adhesive-aluminum, as a
        function of the ultimate temperature of joint formation;
        (●) polypropylene as received; (O) subjected to CASING in
        N_2O for 30 min.

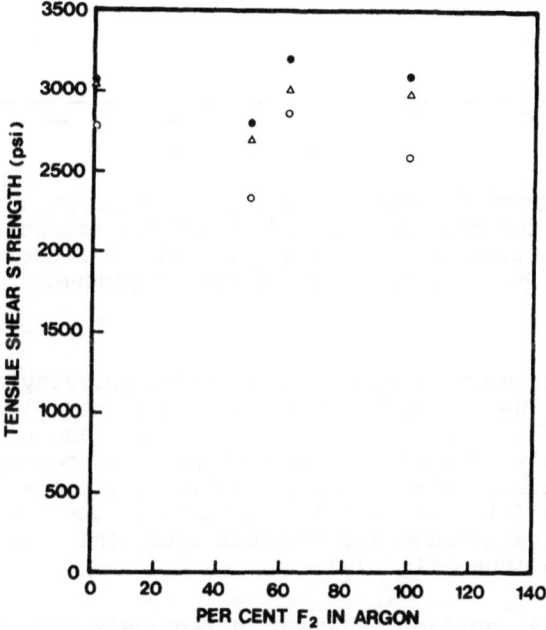

Fig. 13   Tensile shear strength of the composite aluminum-epoxy
          adhesive-fluorinated polyethylene-epoxy adhesive-aluminum
          plotted as a function of volume per cent of fluorine in
          argon.   Polyethylene was exposed to these mixtures for a
          period of 1 hr: (O) 60°C cure temperature; (△) 82°C cure
          temperature; (●) 104°C cure temperature.

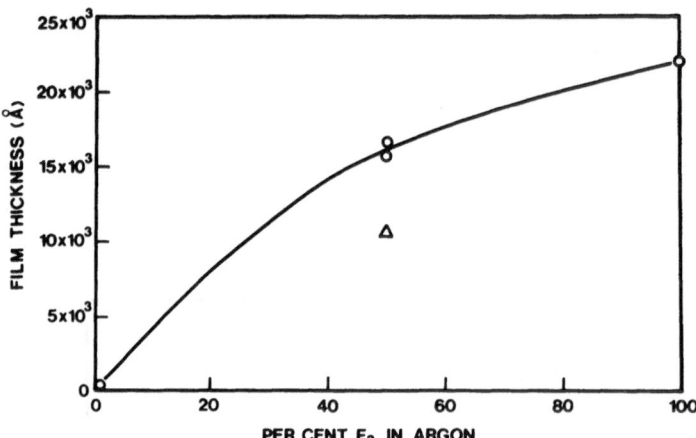

Fig. 14 Estimated thickness of crosslinked region as determined
         from the gel fraction after Soxhlet extraction plotted as
         a function of the volume per cent of fluorine in argon:
         (O) 1 hr. exposure; (Δ) 10 min. exposure.

accomplished for example simply by exposing polyethylene to
fluorine gas. The hydrogen atoms are replaced by fluorine atoms
to generate a fluorinated surface, but at the same time, one also
develops a highly cross-linked network which increases the
cohesive strength of the surface, enabling the preparation of a
strong adhesive joint (Fig. 13). In fact, if one takes these
fluorinated polyethylenes and extracts them, one indeed gets a
cross-linked residue (Fig. 14).

To summarize, surface treatment effectively modifies the
surface region by cross-linking, which in turn upgrades the
mechanical strength. The preparation of strong adhesive joints
can be accomplished in several ways:

* surface treatment that makes the surface more polar or
  receptive to the adhesive,

* inducing cross-linking using the CASING technique which
  effectively joins surfaces without changing γ_c, lowering the
  critical surface tension of wetting by, for example,
  fluorination of some polymer surface. Thus it would appear
  that the mechanical properties of the surface region of a
  polymer exposed to these surface treatments may be of
  overriding importance in making a strong point.

(iii) Adhesion of Polymer Melts: When a polymer melt is in
contact with a metal oxide, it may take a reasonably long period
of time for the polymer to effectively wet and displace the
adsorbed air, in order to form a strong joint. It is well known

that polyethylene is a good adhesive, e.g. polyethylene melt forms
a strong adhesive joint with aluminum oxide. The question is why
does a polymer, when used in the melt form make a strong joint,
and yet the polymer film that we have provided to us by a
manufacturer or that we prepare ourselves, requires a surface
treatment? What is probable is that the way in which you make a
surface is important in dictating whether or not the surface
region requires a surface treatment. Melting a polymer onto a
solid generates a liquid-solid interface, which eventually forms a
solid-solid interface. In this case, the polymer which is a
crystallizable material is being adsorbed on the surface of the
metal oxide and changes its mode of nucleation and crystallization
which is influenced by the substrate. In some way, the substrate
which may be a metal oxide, influences the organization of the
polymer at the solid-liquid interface. A polymer which is
extruded (liquid-vapor interface) eventually forms a film (solid-
vapor interface). Apparently, something is different in the way
you form the surface of the free polymer, as opposed to when the
polymer is formed in contact with a metal oxide.

(iv) Polymer Surface Morphology: If one sections a surface of
polymer formed in contact with a metal oxide, one observes a
completely different morphology (Fig. 15), due to the metal

Fig. 15   Transcrystalline growth of polyethylene adjacent to the
          polymer metal interface.

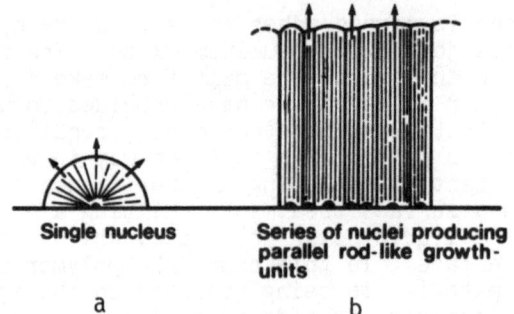

Single nucleus   Series of nuclei producing
                        parallel rod-like growth-
                        units

a   b

Fig. 16 Model for transcrystalline region: (a) bulk nucleation
         and crystallization, no transcrystalline region generated
         at the interface; (b) nucleation and crystallization of
         polymer melt at solid-liquid interface resulting in
         transcrystalline region.

oxide surface. One observes, for example, in polyethylene a
transcrystallization phenomenon. Rod-like growth occurs which
emanates from the surface of the metal oxide (Fig. 16). It is
postulated that in the crystallization process whatever species
are responsible for the mechanically weak surface that exists on
the polymer surface, no longer reside at the solid-solid
interface, but are rejected into the bulk. As a result the
interfacial region of the polymer no longer requires a surface
treatment. To emphasize this point consider polyethylene melted
onto aluminum oxide. Upon removal of the nucleating substrate,
i.e. dissolution of the aluminum, one can form a strong joint to
the polyethylene surface without a surface treatment, using a
conventional adhesive (Fig. 17).

     Thus it seems that the way you make a surface is extremely
important in terms of whether or not you will subsequently have to
surface treat that polymer. In principle, one does not have to
prepare a polymer surface by solidification in contact with a
vapor, as in the extrusion process, but alternatively, one can
form an interface by solidifying the polymer in contact with a
thin metal foil. The need for specific surface treatments
generally results therefore from insufficient knowledge concerning
how a surface region is formed.

     (v) Metallized Polymers: Many polymers can be metallized by
evaporating metals onto them, to make strong adhesive joints, e.g.
decorative plastics used in the automotive and appliance
industries. Why do these metals adhere so well to the plastics,
when, for example, if you wanted to join the polymer with
adhesives, they will not support a strong joint unless surface

treated. A good example is polyethylene which can be metallized
with a variety of metals, forming a strong joint. It appears that
the act of metallization itself in some way modifies the surface
region of the polymer, to form a strong boundary layer. This is
illustrated in Tables 1-3 for Teflon FEP, Teflon TFE and
polyethylene. For polyethylene, when either aluminum or titanium
are etched from the polymer, the presence of a cross-linked layer
is detected upon extracting the polymer with xylene.

Metallization, in some way generates species which lead to a
cross-linking of the surface region of the polymer, thereby
enabling the formation of a strong ahesive joint. A suggested
mechanism for surface cross-linking using conventional adhesives,
is shown in Fig. 18.

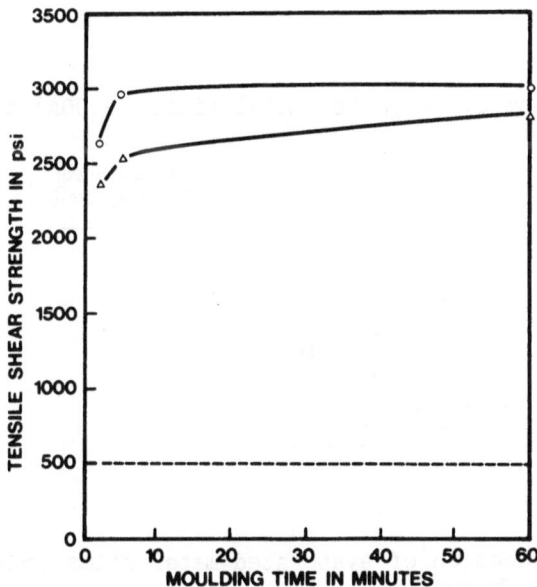

Fig. 17   The tensile shear strength of the composite aluminum-
          epoxy adhesive-polyethylene film-epoxy adhesive-aluminum
          cured at 82°C plotted as a function of molding time at
          175°C.

          △ - Polyethylene film nucleated and solidified in contact
              with etched aluminum surface (subsequently dissolved from
              substrate)

          O - Polyethylene film nucleated and solidified in contact
              with gold (subsequently dissolved from substate)

          (---) - Molded against low energy surface (PTFE or Mylar)

Table 1   Adhesion of evaporated metal films (~60A) to Teflon FEP.

Film	Tensile Shear Strength (kg/cm^2)
1. Teflon FEP (as received, no metallization)	~0
2. Teflon FEP (~60A Au deposited by evaporation)	~0
3. Teflon FEP (~60A Al deposited by evaporation)	94
4. Teflon FEP (~60A Ti deposited by evaporation)	142.5

Table 2   Adhesion of evaported metal films (~100A) to Teflon TFE

Film	Tensile Shear Strength (kg/cm^2)
1. Teflon TFE (as received, no metallization)	~0
2. Teflon TFE (~100A Ti deposited by evaporation)	61.3
3. Teflon TFE (CASING treated in helium for 1/2 hr)	56.0

Table 3   Adhesion of evaporated metal films (~100A) to polyethylene

Film	Tensile Shear Strength (kg/cm^2)
1. Polyethylene (as received, no metallization)	9.3
2. Polyethylene (~100A Ti deposited by evaporation)	115
3. Polyethylene (CASING treated in helium for 10 min.)	138

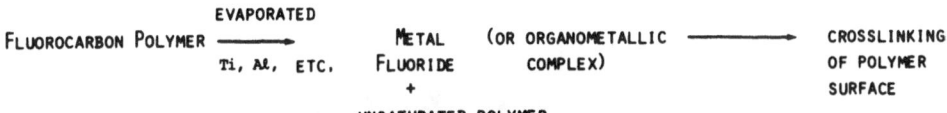

Fig. 18 A schematic of surface crosslinking induced be metal
         evaporation on a fluoropolymer surface.

The mechanism for the generation of a polymer by forming the
catalyst in situ is shown schematically in Fig. 19. This example
may be viewed as a modified Toy process. The wettability may or
may not change, but by and large you are effectively changing the
surface region's structure. Thus the way you make a surface is of
considerable importance because it may dictate whether or not a
surface treatment is required.

(vi) Adhesive Joint Performance: When a bonded composite is
exposed to a hostile environment such as water or a surfactant
solution, it may or may not survive for extended periods of time.
Consider a composite BC, in the presence of both a vapor phase,
and a liquid phase (Fig. 20). Several years ago, Owens, at
Dupont, studied the interaction of a variety of surfactant
solutions in contact with a composite structure. He showed that
if one calculates the thermodynamic work of adhesion, of that
composite in the liquid environment, as compared to contact with
the liquid vapor, it became negative. This implies that the
composite has a tendency to separate. Thermodynamics does not

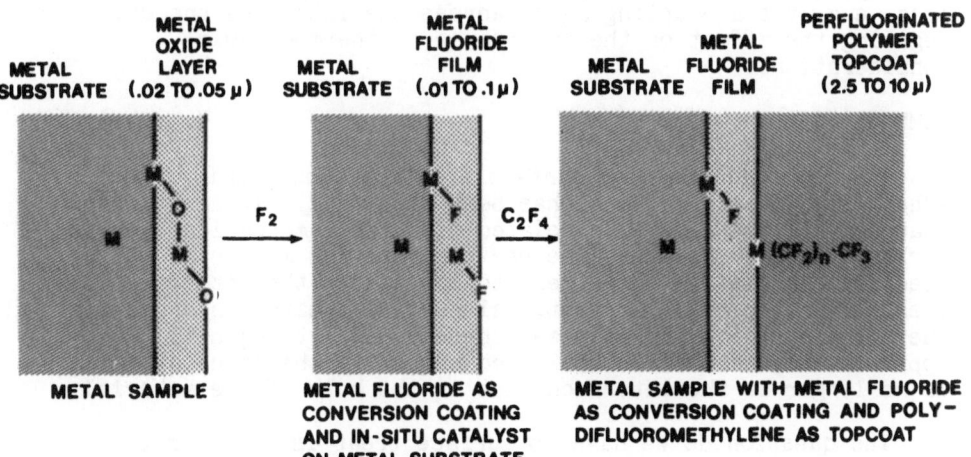

Fig. 19 Schematic of surface reaction of gaseous F_2 on a metal
         surface and the subsequent polymerization of C_2F_4 on the
         activated metal fluoride surface (Toy process).

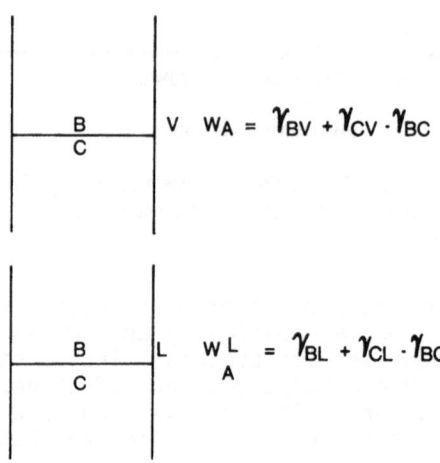

$$W_A = \gamma_{BV} + \gamma_{CV} \cdot \gamma_{BC}$$

$$W_A^L = \gamma_{BL} + \gamma_{CL} \cdot \gamma_{BC}$$

Fig. 20   Schematic of a composite interface in the presence and
absence of liquid. The corresponding works of adhesion
are illustrated

tell you when separation occurs, but it does indicate that there
is a tendency for this laminate to separate. Obviously what is
required is some modification of the surfaces so that the work of
adhesion is no longer negative (with a tendency to delaminate)
but actually becomes positive exhibiting a tendency for it to
remain intact in the presence of a liquid. The driving force, in
developing a permanent composite, would be to generate a covalent
bond, e.g. via a coupling agent across the interface thereby
negating the affect or the importance of thermodynamic
considerations.

SUMMARY

     The important role of surface treatments on joining and
adhesively bonding polymers has been discussed. It is believed
that many polymers have associated with them, mechanically weak
surface layers. However, whether they are low molecular weight
species or additives which tend to migrate to the interface is
presently not clear. Conventional surface treatments cross-link
the surface region and increase the cohesive strength to
approximately that of the bulk, consequently the locus of failure
is no longer at the interface, but in the bulk of the material.

     The question do we have to surface treat is valid in a
practical sense. From a purely technical point of view, however,
you do not have to surface treat. If a surface is prepared in a
completely different manner, one can effectively make a strong
joint, and obviate the use or need for specific adhesives. By a
careful choice of surface treatment, one can generate a composite

structure that will have permanence with respect to a particular environment.

Bibliography

1. H. Schonhorn and L.H. Sharpe, J. Polymer Sci. [B] $\underline{2}$,719 (1964).
2. L.H. Sharpe and H. Schonhorn, in Contact Angle, Wettability and Adhesion, No. 43 in Advances in Chemistry Series, American Chemical Society, Washington, D.C., 1964, p. 18.
3. H. Schonhorn and L.H. Sharpe, J. Polymer Sci. [A] $\underline{3}$,3087 (1965).
4. W.A. Zisman, in Ref, 2, p.1.
5. H. Schonhorn, H.L. Frisch and G.L. Gaines, Polymer Engr. Sci. $\underline{17}$, 440 (1977).

General

R. Houwink and G. Salomon, eds., Adhesion and Adhesives, 2nd ed., Elsevier Publishing Co., New York, 1965.
D.E. Eley, ed., Adhesion, Oxford University Press, 1961.
J.J. Bikerman, The Science of Adhesive Joints, 2nd ed., Academic Press, New York, 1968.
I. Skeist, ed., Handbook of Adhesives, Reinhold, New York, 1962.
P. Weiss, ed., Adhesion and Cohesion, Elsevier, Amsterdam, 1962.
F.P. Bowden and D. Tabor, The Friction and Lubrication of Solids, Part I, 1st ed., 1950; Part II, 1st ed., 1964; Clarendon Press, Oxford University Press.
Contact Angle, Wettability and Adhesion, No. 43, in Advances in Chemistry Series, American Chemical Society, Washington, D.C., 1964.
R.L. Patrick, ed., Treatise on Adhesives and Adhesion, Marcel Dekker, New York, 1966.
A. Sharples, Polymer Crystallization, St. Martin's Press, New York, 1966.

EVALUATION OF FIBER ADHESION IN COMPOSITES

A. T. DiBenedetto

Institute of Materials Science
University of Connecticut
Storrs, Connecticut 06268

INTRODUCTION

Over the last few decades there has been a continuing interest
in the development of composite materials of plastics reinforced
with high strength fibers, such as glass and graphite. Among the
many reasons for this interest is the possibility of fabricating
materials for use as primary and secondary structures that, on the
basis of weight, are stronger and more rigid than conventional
materials of construction.

For example, a plastic reinforced with fifty percent by volume
of a continuous, high modulus graphite fiber, oriented in one direction, can attain a modulus of elasticity and a tensile strength
of the same order as those of a high strength steel, at least in
the direction of the oriented fibers.

Since such a composite weighs approximately one-fifth that of
steel, the specific properties, that is to say the modulus of
elasticity and strength per unit of specific gravity, are five
time greater for the composite material.

All fiber reinforced composite materials, however, are
anisotropic at some level of examination. Even if the fibers are
randomly oriented so that the material is isotropic on a macroscopic scale, locally there is anisotropy in the regions around
individual fibers. Strength and rigidity are high in a direction
parallel to a fiber axis, but perpendicular to that axis the properties fall to the level of the plastic. The angular dependence
of properties is strongly dependent upon the adhesion between the

two phases. Generally, in order to promote high performance and long term stability, it is necessary to form high strength, chemically stable interfaces between the fiber and plastic components (1).

Because of the need to transfer stress from the matrix to the reinforcement, the reinforcing phase must be present with a relatively high ratio of fiber length to diameter. High strength, high modulus reinforcing fibers such as glass or graphite, however, are brittle materials which are susceptible to severe property deterioration and breakage due to surface damage during handling and composite manufacture.

In order both to protect the fibers and to promote high fiber-matrix adhesion, mixtures of sizing and coupling agents are applied to the fiber surfaces during the fiber manufacturing process. The surface treatment of glass fibers is well developed commercially and has been studied at least the past 40 years.

The function of a <u>sizing agent</u> is to protect the fiber from surface abrasion caused by handling and fiber-fiber contacts. For example, a glass fiber might be coated with a thin polymeric film of polyvinylacetate (PVA). The PVA is a tough, ductile, water resistent coating at room temperature thus minimizing surface abrasion and protecting against the damaging effects of humidity. At temperatures that might be encountered in a composites manufacturing process, polymer flow causes a buildup of shear stresses that also cause fiber breakage. Frequently, a mixture of a sizing agent and a low molecular weight lubricant is used to minimize the fiber breakage.

The function of a <u>coupling agent</u> is primarily to promote fiber-matrix adhesion, but when used in sufficient quantity it also serves as a protective coating. A coupling agent is any substance which can interact chemically or physically with both the fiber surface and the polymer matrix. Ideally, one would like to modify the fiber surface so that the polymer matrix can attach itself to the surface by chemical bonding. The most effective and widely used coupling agents for glass fibers are the organosilanes having the general structure, X_3SiR. The R group is chosen to be resinophilic. It may be vinyl, γ-methacryloxy-propyl, γ-amino propyl, etc., depending upon the reactivity of the polymer phase. The X-groups are usually methoxy, ethoxy or chloro. The X-groups hydrolyze in the presence of water to form silanols which in turn can either condense with the silanols present on the glass surface to form siloxane linkages with the glass or can condense with each other to form a siloxane polymer coating on the surface. It is thought that the first molecular layer of silane does the former,

while more distant molecular layers, when present, inter-condense.
Interactions other than chemical reaction can also occur between
the coupling agent and the surface. For example, silane molecules
can also hydrogen bond to the surface, rather than covalently bond.
One theory suggests a mechanism that involves a reversible silanol
bond formation and a dynamic equilibrium of the coupling agent com-
peting with small penetrant molecules such as water (2).

Typical commercial fiber treatments employ mixtures of coupling
agents, sizing agents and lubricants, ordinarily applied in such a
way that the overall weight of the coating is in the range of 0.2
to 2.0 percent of the fiber weight. The choice of components is
specific to the fiber-polymer combination being considered and
optimization of the coating is complicated by the fact that individual components often perform more than one function and in
many instances interfere with each other. The traditional method
of evaluating a composite material by mechanical testing under
various conditions of temperature and environment measures only
the overall fiber reinforcement efficiency and is not capable of
distinguishing the effects of adhesion vis-a-vis improved fiber
protection.

Recognizing the need for improved fiber treatment evaluation
procedures, Ongchin et al. (3) and then Fraser et al. (4) developed an experimental technique and theoretical analysis that
allowed the evaluation of both the adhesion promoting capability,
(or more correctly the stress transfer capability) and the fiber
protection performance of a given fiber treatment formulation.
In the former work (3) the analysis did not take account of either
the strength characteristics of glass fibers or the statistical
nature of fiber fracture in composites, thus giving rise to qualitative estimates of the two factors, while in the latter work (4)
the theoretical analysis was too tedious to have practical value.

In this article, the simple experiment proposed by
Ongchin et al. and Fraser et al. is described and the results of
such a test are used to calculate the value of a parameter, τ ,
that measures the stress-transfer capability of a given fiber
treatment when the fiber is completely encapsulated in a polymer
matrix. The method also permits the evaluation of fiber protection
by measuring the strength - length characteristics of a given
fiber. This procedure may be used for the routine screening and
evaluation of new fiber surface treatments and, furthermore, by
correlating the measured parameters with other composite properties,
such as strength, toughness, impact strength and fatigue resistance,
one may more reliably optimize these properties in a given composite
material. Experimental evidence will be presented that illustrates
the efficacy of the proposed technique.

THE FIBER FRAGMENTATION TEST: A MEASURE OF ADHESION

A single fiber encapsulated in a compression molded test specimen in the form of a dumbell is used in the experimental test. The
mold is shown in Figure 1. A single fiber is mounted on a fork and
is placed in the cavity of the mold between two sheets of plastic.
The samples (eight per shot) are formed under heat and pressure
and subjected to traction in a tensile-testing machine
to an elongation greater than that required to fracture the fibers.
(The matrix material must have an elongation-to-break greater than
that of the fiber.)

Since the volumetric fraction of fiber material in the sample
is less than the critical value (5), the fiber will break into
many small fragments. The lengths of these fragments are a measure
of the ability of the interface to transmit the imposed stresses
from the matrix to the fiber and, therefore, are a measure of the
adhesion between the two phases.

In order to measure the distribution of fiber fragment lengths,
it is necessary to pyrolize the resin, leaving behind the fiber
fragments. A deformed specimen is therefore placed in a Petrie
dish and pyrolized at 350°C for approximately 12 hours, after which
the carbonized mass is placed in a furnace at 550-600°C for about
2 hours to complete the pyrolosis of the resin. The lengths of
the fiber fragments thus obtained can be measured using a projection
microscope of low magnification. The fragments are classified according to their lengths and a cumulative distribution of lengths
is calculated.

In order to have a sufficient number of fragments to determine
a statistical distribution of fragment lengths (100-200 fragments),
it is necessary to accumulate the fragments obtained from about 10
specimens. The relation between the distribution of fragment
lengths and the ability of the interface to transmit to the fiber
the stresses imposed on the matrix depends on the mechanism of
stress transfer between phases.

Figure 2a shows schematically a fiber implanted in a resin.
The traction stress, applied externally, is transferred to the
fiber by means of a shear stress across the interface. The tensile
stress in the fiber increases from nearly zero at the fiber ends
to a maximum value, limited by the breaking stress of the fiber
(Figure 2b). When this limit is reached, the fiber breaks at its
weakest point. If the fiber, however, is not long enough, the
generated tensile stress cannot reach its maximum value and it is
not possible to break the fiber (Figure 2c).

The fiber length for which the generated tensile stress
reaches a value of the breaking strength of the fiber exactly

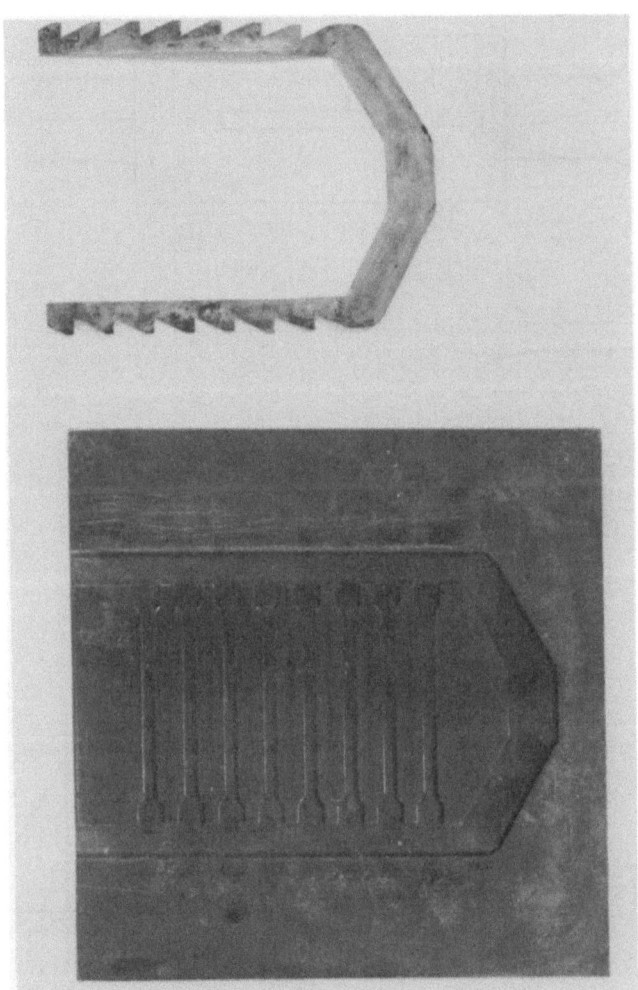

Figure 1. Mold assembly.

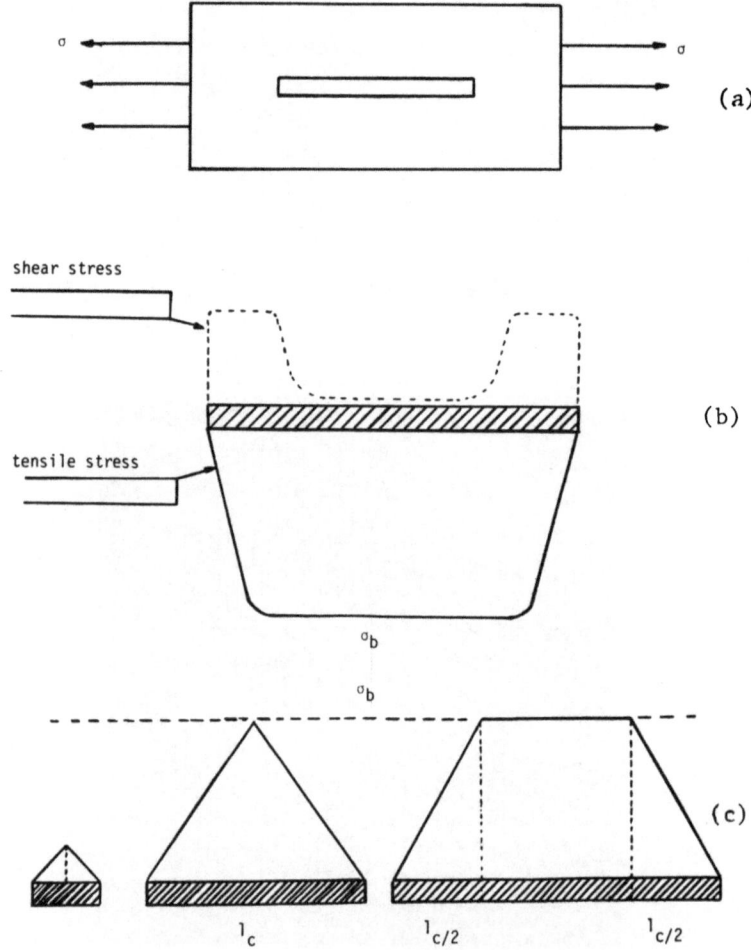

Figure 2.   Stress transfer in a fiber.   (a) Fiber under
            traction; (b) stress distribution in fiber;
            (c) definition of critical length.

at the midpoint of the length (i.e., a triangular stress distribution with the maximum stress equal to the breaking strength),
is called the "critical length", 1_c. Kelly and Tyson (5) have
related the critical length 1_c to the yield stress of the matrix
at the interface τ_y through a force balance around the imbedded
fiber. They assumed that the shear stress at the interface is
constant and equal to the yield strength of the matrix or the
"resistance" of the interface and obtained the following equation
for the critical length:

$$1c = \frac{\sigma_f \; d_f}{2 \; \tau_y} \tag{1}$$

where σ_f is the average breaking strength of the fiber, d_f is
fiber diameter and τ_y is the shear strength of the interface. Thus
if one measures the breaking strength of the fiber, one can estimate an effective interfacial strength τ_y from the critical length
1_c.

When a specimen containing a single fiber is subjected to
tensile deformation and the plastic matrix deforms beyond the
elongation-to-break of the fiber, the fiber will break at its
weakest point as illustrated schematically in Figure 3. At this
point one has two pieces of fiber of different lengths. Under
continued deformation, the fiber stress continues to increase until
it reaches the breaking stress of one of the two fragments, at
which point a second fracture occurs. This process will continue
until all of the fragments are smaller than their critical lengths,
after which fiber fracture is no longer possible. When a fiber of
length 1_c is fractured, two pieces of length $1_c/2$ should be
obtained. The Kelly-Tyson model predicts that the fiber fracture
process should result in a distribution of fragment lengths from
$1_c/2$ to 1_c. If the tensile strength of the fiber is independent
of fiber length, one may calculate an effective interfacial shear
strength τ_y from the measured critical length, using equation 1.

Often, the distribution of fragment lengths obtained is
broader than the 2:1 ratio predicted by the Kelly-Tyson model
(3,4,6). This has been atrributed to the variation of breaking
strength with fiber length. The effect of this variation on the
distribution of fragment lengths is complex and has been reported
in the literature (4).

A reasonable estimate of an "effective" interfacial strength,
τ_e, may be obtained from the midpoint of the cumulative distribution curve (i.e., at $P = 0.5$) where the experimental fragment
length is approximately $3/4$ (1_c). Using equation (1), one may
estimate a value of τ_e from the medium critical length $(1_c)_m$ and
the mean tensile strength at that length, $(\sigma)_m$. In a typical

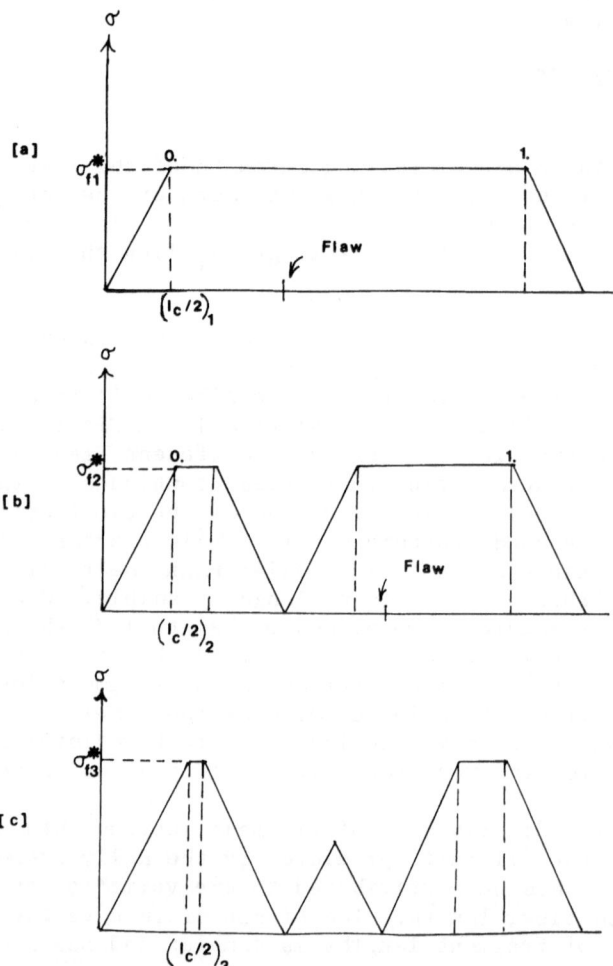

Figure 3.   Stochastic fragmentation model.

composite material, the median critical length may be of the order
of 0.5 to 5.0 mm.

Generally, the tensile strength of fibers are known at lengths
greater than the typical values of $(l_c)m$. It is necessary, therefore, to estimate the tensile strength from data at larger lengths.
Although the tensile strength of brittle fibers, such as glass
fibers, increases bilogarithmically with length (7), over a restricted range of lengths it is often possible to estimate the
strength-length relationship from a simple Weibull function (8).
The resulting strength-length relation is:

$$<\sigma>_1 = \sigma_0 \; \Gamma\left(\frac{1+\alpha}{\alpha}\right)\Big/ l_1^{1/\alpha} \tag{2}$$

where $<\sigma_1>$ is the average tensile strength of a fiber of length
l_1, α and σ_0 are constants and $\Gamma\left(\dfrac{1+\alpha}{\alpha}\right)$ is a gamma function.

Assuming that σ_0 and α are independent of length, one may obtain:

$$<\sigma>_2 = <\sigma>_1 \left(\frac{l_1}{l_2}\right)^{1/\alpha} \tag{3}$$

where $<\sigma>_2$ and $<\sigma>_1$, and the average tensile strengths at lengths
l_2 and l_1 respectively.

Therefore, with data on the cumulative distribution of fragment lengths and on the cumulative distribution of fiber tensile
strengths at one length, an effective interfacial strength can be
estimated using equations (1) and (3).

ANALYSIS OF EXPERIMENTAL DATA

In order to illustrate the experimental technique, the results for three thermoplastic resins reinforced with E-glass
fibers are presented.

The resins used were Nylon 6 (Plaskon 8201), polypropylene
(Shell 5524) and acid polypropylene (Hercules PP-PCO-72). The
fibers were prepared by Owens Corning Fiberglass Company in the
form of multi-filament yarns. The properties and surface treatments are indicated in Table 1.

The distribution of tensile strengths of single fibers of
length 6.35 mm., taken at random from the yarn, was determined
using ASTM D 3379-75. The results of 75 duplicate measurements
were plotted as a Weibull function and the values of σ_0 and α ,
reported in Table 1, were calculated.

Table 1 - Properties of E-Glass Fibers

Type	Treatment*	Diameter d_f (micron)	$<\sigma>_1$ MN/m^2 l_1 = 6,35mm	$\sigma_{\bullet}$ MN/m^2 at l_1 = 6,35mm	α
OCF885	1.6% (by wt.) A-1100/PVA/LUB	14.7	2103	2311	4.38
E-Glass	3.0% (by wt.) A - 1100	11.4	2497	2680	6.59

* A-1100 = γ - aminopropyltriethoxysilane (U.C.C.)

PVA   =   polyvinylacetate

LUB   =   Cerrasol 185A CH_3 $(CH_2)_8$ $CONH_2$

Table 2 - Summary of Results

Fiber and Treatment	Resin Matrix	Shear Strength of Resin τ_y MN/m^2	τ_e MN/m^2
OCF 885 A-1100/PVA/LUB 1,6% by wt.	Nylon-6 Polypropylene	48.3 18.6	34.4 7.1
E-Glass A-1100 3% by wt.	Polypropylene Acid polypropylene	18.6 20.7	7.3 21.0

The distribution of fragment lengths was measured using the technique described in the preceding section. The molding temperatures were 250°C for the Nylon 6 specimens and 175°C for the specimens made from the two types of polypropylene. The molding pressure was 3500 KN/m^2. All specimens were stored in a vacuum oven at 60°C in order to minimize problems associated with water absorption. Test specimens 76 mm. long and with a cross section 6.35 mm. x 1.91 mm. were subjected to tensile deformation at a velocity of 5 mm/min. and then were pyrolized, as described previously. The fiber fragments were recovered and were measured using a low magnification Nikon projection microscope.

The cumulative distributions of the critical aspect ratios, l_c/d_f, are shown in Figures 4 and 5. The cumulative distribution of fragments in the Nylon-6 matrix (Figure 4) is lower and narrower than that in the polypropylene matrix, indicating a greater ability to transfer stresses across the Nylon 6/OCF 885 interfaces. The median value of the critical length is about $l_c = 0.74$ mm. (or $l_c/d_f = 50$) in the Nylon-6 compared to $l_c = 2.65$ mm ($l_c/d_f = 180$) in the polypropylene. The tensile strength of the fiber is approximately 3435 MN/m^2 at $l_c = 0.74$ mm. and 2567 MN/m^2 at $l_c = 2.65$ mm. (from equation 3). The effective interfacial strength is thus 34.4 MN/m^2 for the Nylon-6 (corresponding to about 75 percent of the shear yield stress of Nylon 6) and 7.1 MN/m^2 for the polypropylene (corresponding to about 33 percent of the shear yield stress of polypropylene). Thus the adhesion at the Nylon 6 interface is high, while the adhesion at the polypropylene interface is poor, probably manifesting only frictional forces in the latter case.

The cumulative distributions of fragments in polypropylene and in acid polypropylene, reinforced with A-1100 treated E-glass fibers, are compared in Figure 5. The cumulative distribution for the polypropylene is nearly the same as that in Figure 4 with a median critical length of about $l_c = 2.28$ mm. ($l_c/d_f = 200$). The tensile strength of the fiber at this length is 2917 MN/m^2 and the calculated effective interfacial strength is $= 7.3$ MN/m^2, once again not more than frictional forces at the interface. The addition of acid groups to the polypropylene, however, reduces the dimensions of the fragments and the distribution becomes narrower, indicating an improvement in the adhesion at the interface. The median critical length is reduced to about $l_c = 0.91$ mm. ($l_c/d_f = 80$), the tensile strength at this length is 3353 MN/mm^2 and the calculated effective interfacial strength is 21.0 MN/m^2 (very near the yield strength of the resin).

The results are summarized in Tables 1 and 2. From Table 1, one sees that the E-glass fibers treated with A-1100 are stronger and have a narrower distribution of strengths than those treated

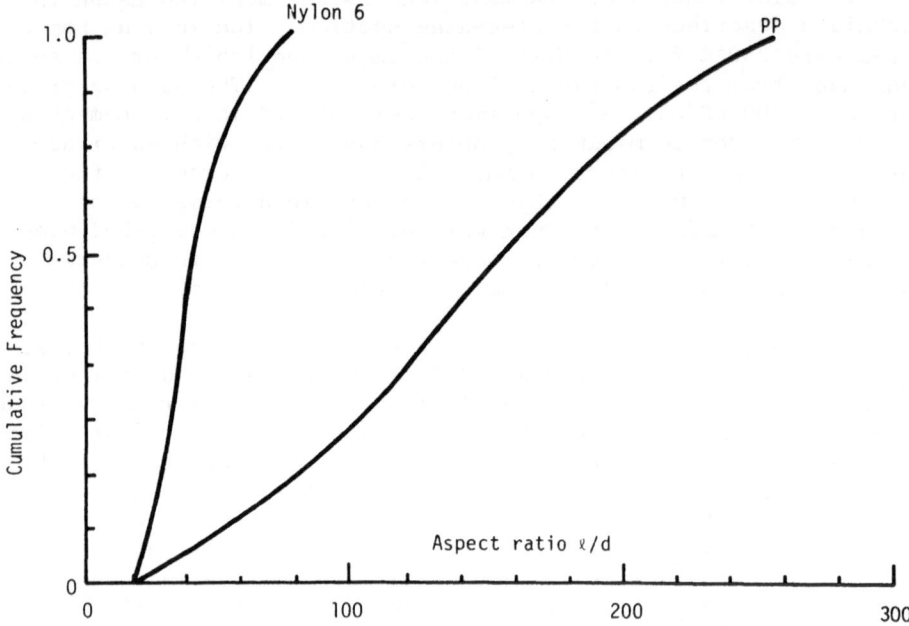

Figure 4.   Distribution of ℓ/d.   OCF885/1.6% (A1100, PVA, LUB)/
Nylon & P.P.

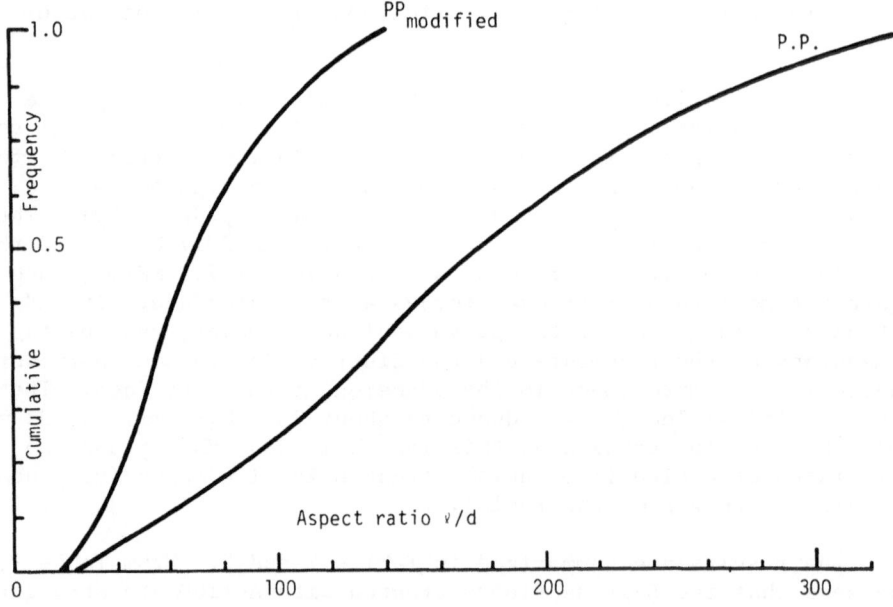

Figure 5.   Distribution of ℓ/d.   E-Glass/3% A-1100/Polypropylene
at 25C.

with the multi-component coating. From Table 2, one sees that the ability to transfer stress (i.e., the interfacial strength) is well developed for both the Nylon 6 and the acid polypropylene matrices, while it is minimal in the propylene matrix.

In conclusion, this technique provides a method for evaluating both the ability of an interface to transmit stresses to the fiber phase and the ability of a surface treatment to protect the fibers from damage that cause a lowering of their tensile strength. It is clear that both high strength fibers and good adhesion between phases are required to develop high strength composite materials.

References

1. W. P. Erickson, E. P. Plueddemann, Chapter 1, Volume 6, Composite Materials Eds. L. J. Broutman and R. H. Krock, Academic Press, New York, (1974).
2. E. P. Plueddemann, Modern Plastics, Vol. 47, 92. (1970).
3. L. Ongchin, W. K. Olender and F. H. Ancker, Proc. 27th Conf. S P I Reinforced Plastics Div., Sec. 11-A (1972).
4. W. A. Fraser, E. H. Ancker, A. T. DiBenedetto, Proc. 30th Conf. S P I Reinforced Plastics Div., Section 22-A, 1-13, (1975).
5. A. Kelly, W. R. Tyson, J. Mech. Phys. Solids Vol. 13, 329 (1965).
6. N. J. Wadsworth, I. Spilling, Brit. J. Appl. Phys. (J. Phys. D.) Ser. 2, Vol. 1, 1049 (1968).
7. B. W. Rosen "Mechanics of Composite Strengthening", Chapter 3 in "Fiber Composite Materials", ASM, Metals Park, Ohio (1965).
8. W. Weibull, "A Statistical Theory of the Strength Materials" Proc. Roy. Swedish Inst. Engr. Res., Ing. Vetenshaps Acad. Handl. Vol. 151, (1939).

BONDING IN WOOD COMPOSITES

J. D. Wellons

Forest Products Department
Oregon State University
Corvallis, Oregon 97331

INTRODUCTION

To form an adhesive bond between wood fibers there must be
flow to bring components in proximity, adhesion across an interface and cure of the adhesive. Because wood fibers are
porous and adsorptive and are a heterogeneous mix of polymers and
oligomers, we seldom can isolate these actions. Thus, I have
chosen to consider the entire bonding process, not merely
adhesion. In fact, experience suggests that adhesion alone is
seldom the reason for bond failure in wood composites.

This paper reviews the state-of-the-art in bonding whole wood
fibers*, individually as in wood fiberboard, or in aggregate as in
laminated lumber or plywood. The focus is on research since 1970,
because earlier work has been carefully reviewed (Marian and
Stumbo, 1962a and b; Jurecic, 1966; Halligan, 1969; Patton, 1970;
Collett, 1972; Venkateswaren, 1975; Wellons, 1977), and on high
yield processes rather than on bonding between cellulose fibers,
as in paper. Further information on bonding in paper may be found
in Stannett, 1967; Weiner, 1977; and Britt, 1979.) Deliberate
attention is given to methods of enhancing the reactivity of fiber
surfaces to form bonds without applying an adhesive.

MECHANISMS FOR ADHESION TO WOOD FIBERS

Our knowledge about mechanisms of adhesion is summarized
briefly. A more detailed review may be found in Wellons (1977).
Although the subject may be divided into mechanical interlocking,

*Fibers that have not been chemically pulped.

127

physical attraction, and chemical bonding, the mechanisms probably occur in combination in real situations.

Mechanical Interlocking

The macroscopic intertwining of adhesive polymer and porous wood fibers occurs when adhesive spreads, penetrates, and wets wood surfaces (Brown et al., 1952). Such interlocking helps transfer stress and thus contributes to bond strength (Horioka, 1973). Most researchers consider this to be a minor bonding mechanism; in fact, some discount it altogether (Salomon, 1965). However, many wood scientists believe that adhesive penetration into the fiber wall on a molecular scale is essential for durable bonds. This penetration has been demonstrated by studies of adsorption of polymer into wood fibers as a function of molecular size (Tarkow et al., 1966; Yutaka and Yasushi, 1977), by light and electron microscopy of gluelines in incident light (Collett, 1970; Nearn, 1974), and by microscopic analysis coupled with infrared (Furuno, 1976), ultraviolet (Nearn, 1974), or x-ray (Smith and Côté, 1971; White et al., 1977) spectroscopy. Nearn (1974) has correlated penetration of fiber walls with bond performance. In spite of a wealth of information supporting the need for fiber wall penetration, it may be unnecessary for adhesion and necessary only to mend submicroscopic fractures (Slaats, 1979) or to provide transition in mechanical properties between the adhesive and wood fiber (White, 1977).

Physical Attraction

Van der Waal's forces and hydrogen bonding (specific adhesion) have long been considered major factors in adhesion in wood composites. Research over the last 20 years documents clearly the relevance of "wetting" (Baier et al., 1968; Marian and Stumbo 1962b; Collett, 1972) and polymer adsorption from solution or melt onto fiber walls. The initial studies of polymer adsorption on wood fibers by Tarkow and Southerland, (1964) have been extended by Okuro (1970), Mizumachi and Fujino (1972), and Mizumachi and Kamidohzono (1975). Immobilization of adhesive polymer by adsorption on wood surfaces has been established beyond doubt, but we have yet to prove the relative importance of such physical attractions to bonding.

Chemical Bonding

The sharing of electrons in covalent bonds between wood fiber and adhesive is probable. Several studies have used hydrolytic rates of wood-adhesive composites (Troughton, 1969; Troughton and Chow, 1968), differential thermal and thermogravimetric analyses

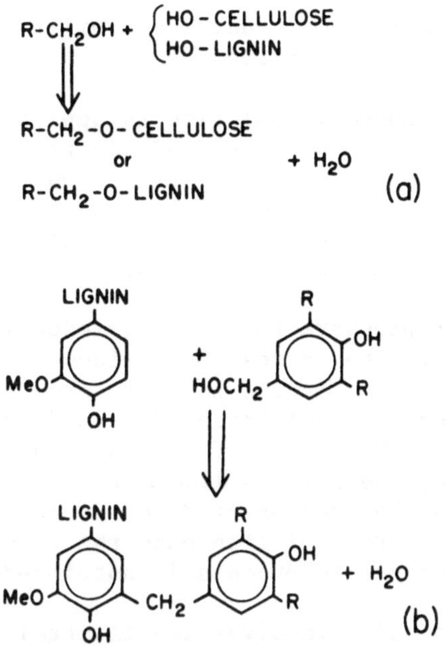
R-CH2OH + HO-CELLULOSE HO-LIGNIN
R-CH2-O-CELLULOSE
or
R-CH2-O-LIGNIN
+ H2O
(a)
LIGNIN
MeO
OH
+
HOCH2
R
OH
R
LIGNIN
R
OH
MeO
CH2
OH
R
+ H2O
(b)

Figure 1. Possible bonding reactions between methylolated adhesive prepolymer and (a) hydroxy groups on wood components or (b) orthohydrogens in lignin.

of similar composites (Ramiah and Troughton 1970; Mizumachi and Morita, 1975), and polymerization kinetics (Chow, 1969) to support the presence of covalent bonds between wood and adhesive. The studies propose a bonding mechanism such as that shown in Figure 1a. Allan and Neogi (1971) used model compounds to establish possible reactions of phenolic adhesives with lignin (Figure 1b), but they found no reaction with carbohydrates. These studies have verified that covalent bonds can form in vitro between formaldehyde based resins and wood fibers. Still to be proven is whether such linkages can form under bonding conditions and whether they are essential to the forming of water-proof bonds.

FACTORS AFFECTING BOND FORMATION

The bonding of an adhesive to wood fibers involves thermodynamic and kinetic parameters and their interaction. Consider the gluing of wood flakes or veneer with a liquid phenol-formaldehyde resin. Let's start with the application of adhesive to wood. For flakes or particles, the liquid resin is atomized and sprayed as droplets onto the surface. For veneer, a film of liquid adhesive is roll-coated onto the wood. To give the adhesive polymer mobility, it is applied to wood as a solution, usually in aqueous alkali, but occasionally in water and alcohol

mixtures. In some instances, adhesive mobility is obtained by
heating, rather than by solvation.

     As soon as the adhesive solution is applied to the wood, the
solvent "wets" and then penetrates and adsorbs into the fibers,
even without pressure, leaving most of the phenolic resin solids
on the fiber surface. The rate of the loss of solvent depends on
the moisture content of the wood fibers, roughness of the surface,
and the inherent attraction between wood and solvent.

     Application of pressure to adhesive-laden flakes or veneers
spreads and transfers the increasingly viscous adhesive and causes
bulk flow of adhesive solids and solvent through the wood surface
into the large pores and capillaries in the fiber walls, thus reducing the thickness of the film. If too much adhesive penetratres into
the fiber structure, the film is discontinuous on the fiber surfaces.
Simultaneously with the pressure-induced penetration, adhesive
polymer is adsorbing from solution onto the fiber and adhering--
provided this thermodynamic process is not impeded by viscosity.

     Cure of the adhesive involves two interrelated processes:
solvent loss and chemical crosslinking. A high solvent content
of the adhesive retards crosslinking. Too rapid loss of solvent
makes the immobile solids incapable of forming a crosslinked film.
Heating increases both solvent loss and crosslinking, although
each rate increases differently with temperature.

     Most of the gluing parameters are affected by time,
temperature, pressure, and fiber properties, so we can manipulate
the adhesive properly to form a good bond. The thermodynamic processes of wetting and adsorption are less amenable to manipulation
than other gluing parameters.

Wettability

     The basic concept of surface energy and its influence on
forming an interface have been thoroughly reviewed (Zisman, 1963,
1972; Huntsberger, 1978). Some argue that the Young-Dupré
equation and its variations do not apply to wood fibers because
they are adsorptive and swell; however, the preponderance of
literature on adhesion (Patton, 1970; Collett, 1972) verifies that
the concept is valid regardless of the substrate, though modifications are needed to account for swelling and solvent adsorption.
Recent studies with plastics show one way to apply the concepts of
surface energy to wood fibers (Good and Kotsidas 1979). We must
note though, that good wetting does not assure good bonding. The
assumption that it does accounts for many discrepancies in the
literature on wood gluing.

Many studies have estimated the surface energy of wood fibers. The critical surface tension of various liquids on wood and wood components varies from 20 to 90+ dynes/cm, depending on the liquids and the previous treatment of the wood. Luner and Sandell (1969) and Lee and Luner (1972) used nonpolar organic liquids to estimate critical surface tensions of assorted dry wood components (Table 1). Their low values and limited range show the tendency of wood components to be hydrophobic when dry.

Because most wood is glued with aqueous adhesive systems, more meaningful values are given by studies using aqueous solutions to estimate the critical surface tension (Gray, 1962; Herczeg, 1965; and Nguyen and Johns, 1979). These show that critical sur-face tensions of wood vary considerably. The highest values reported are for freshly prepared surfaces of wood low in extrac-tives content. The presence of nonpolar extractives, aging of the surface, and heating and drying of the wood fibers all decreased surface energy and wetting. Nguyen and Johns (1979) partitioned surface free energy into polar and dispersion components to better describe the effects of age and extractives on wetting. All of these studies suggest that wettability may frequently limit bonding with aqueous adhesives because the wood may have a critical surface tension less than the surface tension of water (72 dynes/cm).

Attempts to relate wetting contact angles to gluability have had mixed success. Wettability and gluability tend to correlate with adhesives containing organic solvents or acidified water. For example, Freeman (1959), Chen (1970), and Elbez (1978) have successfully correlated the bond quality of urea resins to wettability. Elbez also found correlations between the bond quality of both polyvinyl acetate and resorcinol (organic solvent)

Table 1. Critical Surface Tension of Wood Components
          (Luner and Sandell, 1969; Lee and Luner, 1972).

Component	Critical Surface Tension $ergs/cm^2$
Cellulose	35.5–42.5
Hardwood Xylan	34.0–36.5
Softwood Xylan	35
Galactoglucomannan	36.5
Arabinogalactan	33
Lignin	33–37

resins and wettability. But many studies with adhesives dissolved
in aqueous alkali show little or no relationship between gluing
and wettability of wood fibers. With resorcinol resin, Chen
(1970) found no relationship and with alkaline phenolic adhesive,
Bodig (1962) found it necessary to measure wettability on micro-
tomed wood surfaces in order to relate the results to bond
quality.   Jordan and Wellons (1977) found little relationship
between bond quality of alkaline phenolics and wettability, even
though contact angle measurements suggested that wetting would be
minimal; and Hse (1972a) found that bond quality with such phenolics improved as wetting diminished.

     Studies with wood encounter problems that may explain the
difficulty in relating bond quality to wettability. Few have
attempted to cope with roughness of wood fiber surfaces and its
influence on wettability. In addition, contact angles of liquids
depend on moisture content (MC) of the wood. A recent study in
our laboratory (Wellons, 1980), documented that the contact angle
of aqueous caustic (pH = 11) varied from 0° on veneers with more
than 20% MC to as high as 120° on dry wood (Figure 2). No
equilibrium contact angle could be obtained because it decreased
with time as the wood adsorbed the liquid and the MC changed.
When the dry and apparently poorly wetted veneers were
glued with a phenolic adhesive, the adhesive "dried out" due to
excessive wetting, even in the shortest possible assembly time.

     Another recent study in our laboratory illustrates further
the difficulty of obtaining valid wetting measurements for wood.
Kadlec (1980) overheated veneers in a platen dryer in an attempt
to limit wetting during gluing. He found that water had high contact angles on unsanded surfaces (Figure 3), whereas 0.5% aqueous
caustic (surface tension ≅ 72 dynes/cm) wetted better, probably
due to the greater ability of aqueous caustic to swell and dissolve
components on the wood surface. When the wetting liquid was an
aqueous solution of phenol and formaldehyde (surface tension = 43
dynes/cm), it had a 0° contact angle and immediately adsorbed into
the wood. The ability of aqueous phenol to swell and dissolve
wood components is well established (Stamm, 1964). Because of the
high viscosity, contact angle measurements were not reliable with
the alkaline phenolic resin used to glue the veneers. Others have
also encountered this problem (Chen, 1972; Wilson et al., 1979).
Apparently, however, the phenolic resin adequately wets the
veneers, because all of the platen-dried veneers gave excellent
bonds when glued.

     Hse's results (1972b) illustrate the difficulty of measuring
wettability with phenolic resins. He found that contact angle
between phenolic resins and southern pine veneer decreased as
caustic in the adhesive increased (reduced viscosity and increased
power of the adhesive to swell the wood) and that contact angle

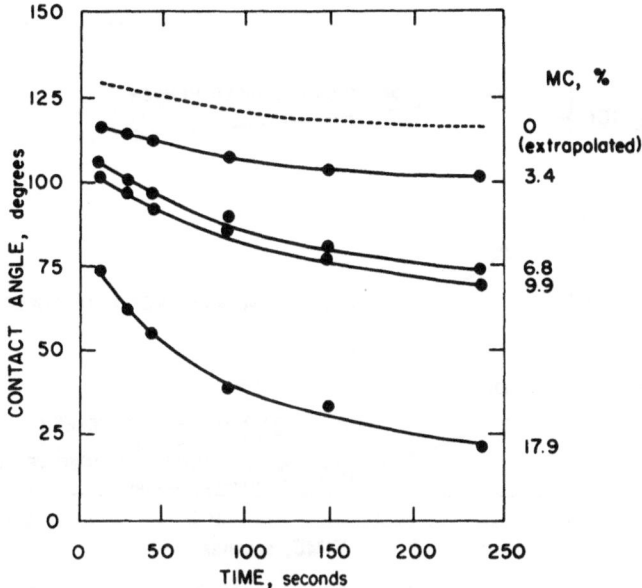

Figure 2.   Effect of time and moisture content on the wetting contact angle of aqueous caustic at pH = 11 (Wellons,
1980).

increased as proportions of formaldehyde increased (higher viscosity due to higher molecular weight and more branching).

     Contact angles have never been measured when adhesives are at
the bonding temperature used for phenolic resins, 100°-150°C
(other liquids have been studied at ambient temperatures), therefore limitations anticipated in wettability may disappear under
actual bonding conditions. The results of these studies do
suggest strongly that poor wetting is rarely a limitation with
aqueous alkaline phenolics.

Wood Properties

     Although many wood properties influence bonding, MC is most
important. As I have indicated, it influences wetting and solvent
loss from the adhesive. When MC is high, wetting with aqueous
adhesives occurs readily, but the amount of water adsorbed is low.
The studies of Northcott et al. (1959, 1962) show clearly the
effect of wood MC on glueline MC. At any specific wood MC, the
assembly time and amount of adhesive must be adjusted for optimum
bond quality. Because wood MC regulates glueline MC, it also
helps regulate adhesive penetration (Villaflor, 1973). The lower
the glueline MC when pressure is applied, the shallower the adhesive penetration. For the same reasons, wood MC influences the
rate of glueline heating and cure (Pillar, 1966; Huynh et al.

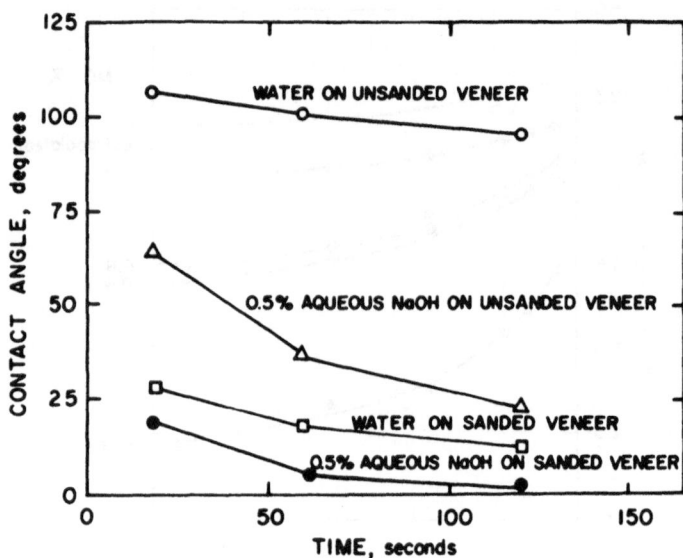

Figure 3.   Contact angles for liquids on 0.3-inch heartwood veneer
platen dried at 425°F (Kadlec, 1980).

1978). The rate of adhesive cure can double when the wood
MC is decreased from 12% to 6%.

Density also affects gluing in several ways. Gluelines may
vary in thickness because dense woods are more difficult for adhesives to penetrate and are less conformable under pressure (Dougal
et al., 1980). Because most structural wood adhesives do not fill
gaps, thick areas in gluelines tend to be weak. Dense woods exert
more stress on cured gluelines because of changes in wood moisture
content (Northcott, 1964). In addition, the greater the wood
density, the greater the compression required to consolidate particles and flakes into a board (Rice and Carey, 1978). The broad
range of densities that may be found within the same piece of wood
may create bonds of different quality in close proximity (Hse,
1968).

Also affecting bonding are the physical characteristics of
the wood surface, such as anatomical roughness, porosity, and
micro-fractures. Wood veneer and particle surfaces are
exceedingly rough (Figure 4) both from anatomical structure and
machining fractures. Additional roughness is probably not
desireable (Marian et al., 1958; Jokerst and Stewart, 1976).

Chemicals composing the wood surface can profoundly affect
bond quality. Carbohydrates have different potential for covalent
bonding than lignin, although both bond well with phenolic and
urea adhesives. However, the high bonding potential of freshly
machined surfaces may be lost with aging (Stumbo, 1964; Nguyen and

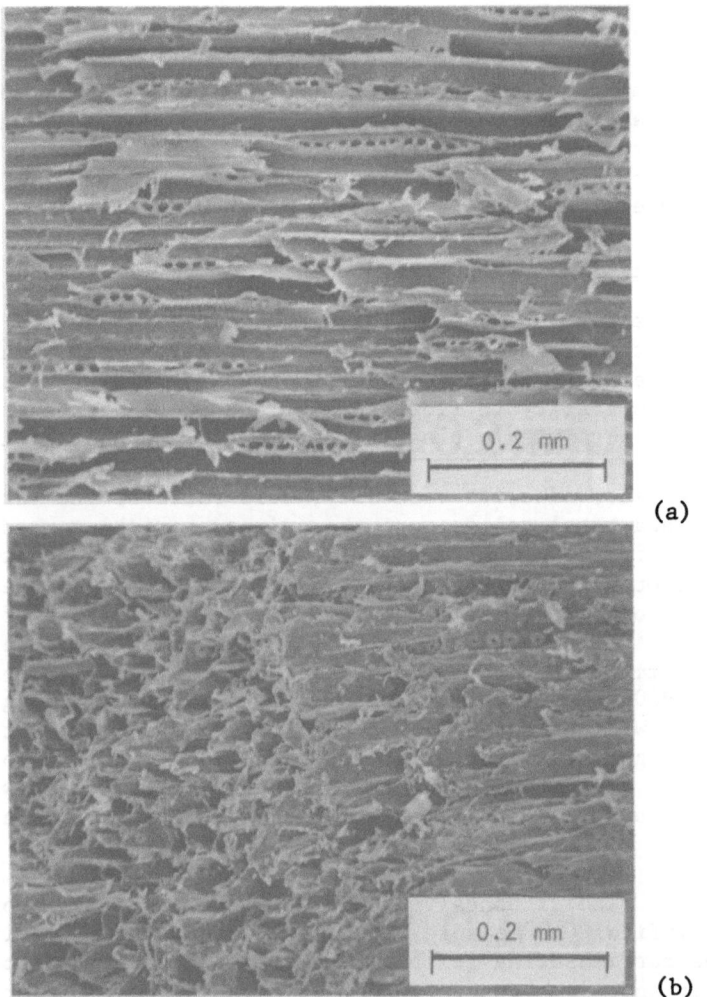

(a)

(b)

Figure 4. Surfaces of Douglas-fir (a) rotary-peeled veneer and
           (b) hammermilled particles.

Johns, 1979) or surface inactivation. The latter phenomena--a
loss of the bonding potential of phenolic adhesives due to
excessive heating of wood surfaces--is critical for wood composites
expected to be durable in exterior use. Further discussion of
this subject may be found in Northcott et al. (1959), Hancock
(1963), Hemingway (1969), Troughton and Chow (1971), and Chow
(1971).

    Extractives on the wood surface influence bonding by reducing
wettability, inhibiting adhesive penetration, retarding or preventing adhesive cure, or diluting the adhesive solids. These
effects are broadly summarized by Plomley et al. (1976). Although

extractives cause few bonding problems with conifers, they are a
frequent cause of problems with hardwoods, such as those attributed to the high concentration of acidic extractives in some
southeast Asian hardwoods (Wellons et al. 1977). Acidic extractives have also created problems in bonding pressure-refined fiber
from North American hardwoods (Albritton and Short, 1979), and
related problems may be more common in the future.

Adhesive Properties

Adhesives for wood composites are discussed elsewhere in this
proceedings, therefore selected adhesive properties are mentioned
here only briefly. Whether an adhesive can perform as needed
depends on its ability to flow, to wet, and to penetrate the fiber
walls, and to cure within prescribed time and temperature limits.

In the past, adhesive viscosity, mole ratio of the various
reactants, and total solids were the primary variables used to
control adhesive flow and penetration (Hse, 1971, 1972a; Steiner,
1973). But other important properties could not be measured.
Little was known about the molecular weight distribution of
resins. With the development of gel permeation chromatography
(GPC), information began to accumulate about molecular sizes
(Armonas, 1970), and the coupling of laser light scattering with
GPC systems (Wellons and Gollob, 1980) now allows us to determine
molecular weight distribution of adhesive resins. Our laboratory
is engaged in a major study on the influence of molecular weight
distribution and flow and penetration of adhesive resins into
wood.

The cure rate of adhesive resins, another adhesive property
that was difficult to measure, crucially affects bond quality.
Procedures for measuring gel time and stroke cure are common in the
literature; and, more recently, differential scanning calorimetry
has allowed evaluation of the relationship of time and temperature
to cure (Chow et al., 1975; Chow and Steiner, 1979a and b). These
studies are establishing cure rates of different adhesives, are
defining threshold temperatures, and are providing insight into
the effects of formulation variables on resin cure.

Adhesive Distribution

No discussion of bonding in wood composites would be complete
without mention of the importance of adhesive distribution.
Neither a resin nor wood property, it nevertheless involves properties of both components.

Unlike laminated lumber and plywood, particle, flake, and
fiberboards are "spot welded" by small amounts of adhesive
(relative to surface area) applied as powder particles or liquid

droplets between the wood particles. Although a continuous adhesive film would undoubtedly be preferable (Marian, 1958), the cost is prohibitive. Thus, we have a continuing debate about measuring and determining the ideal adhesive distribution. It can be qualitatively evaluated by light or electron microscopy, with or without dyes (Lehmann, 1968; Wilson and Krahmer, 1976; Gibson and Krahmer, 1980). Quantitative analysis requires first a decision—whether to measure resin distribution between like-size particles (within-screen fractions) or between different-size particles (between-screen fractions).

Screen fractionation and determination of resin distribution from one particle size to another has assisted the design of blending systems for urea resin. Schwarz et al. (1968), Maloney (1970), Wilson and Hill (1978), and others have determined the urea resin content of various particleboard screen fractions, usually by nitrogen analysis, and have related these values to the mass and surface area available in those fractions. These studies have helped diagnose problems in blending systems due to excessive adhesive capture by either coarse or fine particles. Similar results have not been obtained for phenolic binders because suitable techniques are not available to quantitate cured phenolic resins (Moore, 1975).

Carroll and McVey (1962) argued that a more serious problem of adhesive distribution may occur with particles of like size, some capturing much resin and others none. They showed the potential magnitude of this problem by including particles without adhesive resin in their panels, but they could not prove that normal commercial conditions produce such unequal distribution. Distributions of this type were verified recently by Kasper and Chow (1980) with x-ray spectrometry of individual flakes containing phenolic resin, while Wellons and Wilson* determined nitrogen content of individual particles. Both studies concluded that inefficient blending can reduce the strength of composites because individual flakes or particles receive little or no adhesive.

PROBLEMS AND OPPORTUNITIES

In the coming decade, we will need to solve many problems in bonding wood fiber. Raw material shortages will increase costs, while some economic deterents to new gluing technology will be overcome by governmental regulations. Thus, our scientific skills will be presented with challenging opportunities.

---

*Wellons, J. D. and J. B. Wilson. Resin distribution within particle fractions. Manuscript in preparation for publication.

The Changing Substrate

Wood fiber has become a scarse commodity which forces us to use species that we formerly avoided. Softwoods are being replaced by hardwoods in many products; composites are increasingly made from more dense woods and woods richer in extractives. These changes in the substrate require changes in the bonding system.

The form of the wood also is changing; wet fiber processes are being replaced by dry fiber processes in the fiberboard industry. Structural panels are being made from flakes rather than veneers. Both changes create problems that may make adhesive distribution more important than any other step in the bonding process.

The increasing insistence of building codes that wood fiber products do not decay or burn means we are now asked to modify adhesive systems to cope with large concentrations of oily liquid or organic and inorganic salt. While some of these additives are compatible with traditional adhesives, others totally incapacitate them (e.g. Bergin, 1963; Chen, 1974). Such problems will increase, requiring greater variety in adhesive materials.

Adhesive Supply

The petrochemical industry uses a small portion of our threatened oil and gas reserves. Therefore, though phenol, urea, and methanol will likely be available for many years, price increases may be dramatic (White, 1979). Thus, economics is expected to favor increased use of lignins, tannins, or other "natural" materials in adhesives.

Concern over the possible health effects of formaldehyde vapors may be the most important impact on adhesive supply (Sundin, 1978; Nestler, 1977). Although reformulation of urea and phenolic resins may minimize formaldehyde emissions from the cured resins (Peterson et al. 1974; Minemura et al., 1976; Raffael, 1978), some emission standards currently being discussed would be difficult, if not impossible, to meet. Isocyanates and other adhesive polymers may thus displace currently used adhesives, whether or not they are economically favorable.

Bonding Without Adhesives

One alternative to many bonding problems would be to enhance the tendency of polar polymers in wood fiber to bond without adhesives. If lignin can help bind carbohydrate microfibrils into solid pieces of wood, can it not be used as a binder in reconstituted products? Lignin has been a binding agent for wet-process fiberboard for many years (Spalt, 1977), and bonding has been enhanced by adding acid salts and oxidants such as sulfuric acid, aluminum sulfate, and ferric sulfate. Hydrogen bonding between carbodydrates from fiber walls largely determines the bonding of paper fibers and the science of paper physics has developed add- itives to enhance these bonds. Recent research has shown that heat and pressure can consolidate dry pulp fiber, achieving thermo- plastic flow of hemicellulose and lignin (Bach et al., 1979;Byrd, 1979; Horn, 1979). But these processes result in lower energy bonds that are not durable enough for many composites.

In recent years, much research has focused on increasing the possibility of covalent bonds to the wood surface. Several scien- tists have examined difunctional monomers capable of reacting with hydroxyl groups on the wood surface. Schoring et al., (1972) used hexamethylene diamine to link wood components to polyvinyl chloride. Collett (1973) used nitric acid to oxidize the wood surface before reacting it with hexamethylene diamine. He hoped that the diamine would bridge fibers and particles in his composites. Excellent bonds were obtained, but only after including 10% diamine and consolidating the product to specific gravities of 0.8 or greater. Pohlman (1974) used dibasic acids with catalysts to bond wood particles. Maleic anhydride and ferric chloride combined to provide bonds equivalent to phenol-formaldehyde resins, but high specific gravities were required, and the residual acid may lead to bond degradation with time.

Several oxidative treatments have enhanced bonding. Goring and Suranyi (1969) and Kim and Goring (1971) used ozone to create carboxyl groups on cellulose fiber, wood, and plastic surfaces, creating better bonds between these various constituents. Young (1978) used dichromate in oxalic acid as an oxidant to enhance bonding between cellulose fibers. Allan et al. (1971) studied the oxidative coupling of phenolics to lignin-containing fibers in the presence of strong oxidants like potassium ferricyanide and ferric chloride. Stofko (1972) consolidated wood flour and various wood com- ponents by combining hydrogen peroxide with sodium chlorite and ferrous sulfate to catalyze oxidative coupling reactions. Johns and Nguyen (1977) and Johns and Woo (1978) further developed Stofko's system by forming peroxyacids on the wood or fiber sur- face before consolidation with heat and pressure. Brink et al. (1977) give details of many of these studies. Although most of these procedures are thought to generate covalent bonds to wood fiber surfaces, these simple reagents are unable to bridge the large voids between neighboring fibers. Though the technology may be applicable to high density fiber products, its application to lower density particle, flake, and veneer composites is more difficult.

To achieve "closeness", polymers or resins were needed for bridging. This led Johns et al. (1978) and Brink et al. (1980) to

use lignosulfonates and furanes with nitric acid and peroxyacids
to allow extension of the oxidative coupling concept to lower density composites. Research in this area is still in infancy but is
proceeding empirically to define ranges of application. Although
these bonding systems are beginning to look increasingly like traditional adhesives, they may ultimately provide needed clues for
ways to form durable adhesives with lignins and tannins.

LITERATURE CITED

Albritton, R. O. and P. H. Short, 1979, Effects of extractives
    from pressure-refined hardwood fiber on the gel time of urea-
    formaldehyde resin, For. Prod. J. 29(2):40-41.
Allan, G. G., P. Mauranen, A. N. Neogi, and C. E. Peet, 1971,
    Grafting of phenolic compounds onto lignocellulosic fibers
    by oxidative coupling, Tappi 54(2):206-211.
Allan, G. G. and A. N. Neogi, 1971, Mechanism of adhesion of
    phenol-formaldehyde resins to cellulosic and lignocellulosic
    substrates, J. Adhes. 3(1):13-18.
Armonas, J., 1970, Gel permeation chromatography and its use in
    the development of resins, For. Prod. J. 20(7):22-28.
Bach, E.L., S. Christer, G. Walstrom, and R. G. Anderson, 1979,
    Bonding in paper webs under water deficient conditions, Tappi
    62(3):89-92.
Baier, R. E., E. G. Shafrin, and W. A. Zisman, 1968, Adhesion:
    mechanisms that assist or impede it, Science 162:1360-1368.
Bergin, E. G., 1963, The gluability of fire-retardant treated
    wood, For. Prod. J. 13(12):549-556.
Bodig, J., 1962, Wettability related to gluability of five Philip-
    pine mahoganies, For. Prod. J. 22(6):265-270.
Brink, D. L., B. M. Collett, A. A. Pohlman, A. F. Wong, and J.
    Philippou, 1977, Bonding of lignocellulosic surfaces by oxida-
    tive treatment and monomeric or simple polymeric crosslinking
    agents, in: "Wood Technology: Chemical Aspects," I.S.
    Goldstein, ed., Symposium series #43, American Chemical
    Society, Washington, D.C.
Brink, D. L., M. L. Kuo, W. E. Johns, M. J. Birnbach, H. D. Layton,
    and T. Nguyen, 1980, The bonding of lignocellulosic materials
    using oxidative pretreatments and crosslinking agents, For.
    Prod. J. In press.
Britt, K., 1979, Sheet formation—advances in retention and
    drainage concepts. Pulp Pap. Mag. Can. 80(60): T152-T156.
Brown, H. P., A. J. Panshin, and C. C. Forsaith, 1952, "Textbook
    of Wood Technology, Volume II", McGraw Hill Book Co., N.Y.
Byrd, V. L., 1979, Press drying—flow and adhesion of hemicellu-
    lose and lignin. Tappi 62(7):81-84.
Carroll, M. N. and D. T. McVey, 1962, An analysis of resin effi-
    ciency in particleboard, For. Prod. J.:305-310.
Chen, C., 1970. Effect of extractive removal on adhesion and wet-
    tability of some tropical woods, For. Prod. J. 20(1):36-41.

Chen, C., 1972, Measuring the wetting of wood surfaces by
    adhesives, Mokuzai Gakkaishi 18(9):451-456.
Chen, C., 1974, Gluing study of pyresote-treated, fire-retardant
    plywoods, For. Prod. J. 25(2):33-37.
Chow, S., 1969. A kinetic study of the polymerization of phenol-
    formaldehyde resin in the presence of cellulosic materials,
    Wood Sci. 1(4):215-221.
Chow, S., 1971, Infrared spectral characteristics and surface
    inactivation of wood at high temperatures, Wood Sci.
    Technol. 5:27-39.
Chow, S. and P. R. Steiner, 1979a, Comparison of the cure of
    phenol-formaldehyde novalac and resol systems by differential
    scanning calorimetry, J. Appl. Polym. Sci. 23:1973-1985.
Chow, S. and P. R. Steiner, 1979b, Comparison of curing and bonding
    properties of particleboard- and waferboard-type phenolic
    resins, For. Prod. J. 29(11):49-55.
Chow, S., P. R. Steiner, and G. E. Troughton, 1975, Thermal
    reactions of phenol-formaldehyde resins in relation to molar
    ratio and bond quality, Wood Sci. 8(1):343-349.
Collett, B. M., 1970, Scanning electron microscopy: a review and
    report of research in wood science, Wood Fiber 2(2):113-133.
Collett, B. M., 1972, Surface and interfacial adhesion in wood science
    and related fields, Wood Sci. Techn. 6:1-42.
Collett, B. M., 1973, Oxidative mechanisms for polymerization of
    lignocellulosic materials. Ph.D. dissertation, University of
    California, Berkeley.
Dougal, E., R. L. Krahmer, J. D. Wellons, and P. Kanarek, 1980,
    Glueline characteristics and bond durability of southeast
    Asian species after solvent extraction and planing of
    veneers, For. Prod. J. In press.
Elbez, G., 1978, Study of the wettability of wood, Holzforschung
    32(3):82-92.
Freeman, H. A., 1959, Relation between physical and chemical
    properties of wood and adhesion, For. Prod. J.
    19(12):451-458.
Furuno, T., 1976, Structure of the interface between wood and
    synthetic polymer, Mokuzai Gakkaishi 22(8):473-478.
Gibson, M. and R. L. Krahmer, 1980, Staining to make urea-
    formaldehyde resin visible on glued wood surfaces, For.
    Prod. J. 30(1):46-48.
Good, R. J. and E. D. Kotsidas, 1979, Contact angles on swollen
    polymers: the surface energy of crosslinked polystyrene, J.
    Adhes. 10(1):17-24.
Goring, D. A. I. and G. Suranyi, 1969, Improved bonding of cellu-
    lose and ether polymers by surface treatment with a gas,
    Pulp Pap. Mag. Can. 70(10):102-110.
Hancock, W. V., 1963, Effect of heat treatment on the surface of
    Douglas-fir veneer, For. Prod. J. 13(2):81-88.
Hemingway, R. W., 1969, Thermal instability of fats relative to
    surface wettability of yellow birch, Tappi 52(11):2149-2155.

Herczeg, A., 1965, Wettability of wood, For. Prod. J. 15(10):
    499-504.

Horioka, K., 1973, The mechanism and durability of adhesion for
    wooden members, proceeding of IUFRO meeting, Division 5,
    Sept. 22-Oct. 12, Republic of South Africa 2:503-527.

Horn, R. A., 1979, Bonding in press dried sheets from high-yield
    pulps, Tappi 62(7):77-80.

Hse, C. Y., 1968, Gluability of southern pine earlywood and
    latewood, For. Prod. J. 18(12):32-36.

Hse, C. Y., 1971, Properties of phenolic adhesives as related to
    bond quality in southern pine plywood, For. Prod. J.
    21(1):44-52.

Hse, C. Y., 1972a, Influence of resin formulation variables on
    bond quality of southern pine plywood, For. Prod. J.
    22(9):104-108.

Hse, C. Y., 1972b., Wettability of southern pine veneer by phenol
    formaldehyde wood adhesives, For. Prod. J. 22(1):51-56.

Huntsberger, J. R., 1978, Interfacial energies, contact angles
    and adhesion. Adhes. Age 1978 (December):23-27.

Huynh, H. K., P. E. Lancaster, P. Lepoutre, and A. A. Robertson,
    1978, The setting of aqueous adhesive on paper, Tappi
    61(12):63-65.

Johns, W.E., H. D. Layton, T. Nguyen, and J. K. Woo, 1978, Non-
    conventional bonding of white fir flakeboard using nitric
    acid, Holzforschung 32(5):162-166.

Johns, W. E. and J. K. Woo, 1978, Surface treatments for high
    density fiberboard, For. Prod. J. 28(5):42-48.

Johns, W. E. and T. Nguyen, 1977, Peroxyacetic acid bonding of
    wood, For. Prod. J. 27(1):17-23.

Jokerst, R. W. and H. A. Stewart, 1976, Knife- versus abrasive-
    planed wood: quality of adhesive bonds, Wood Fiber
    8(2):107-113.

Jordan, D. L. and J. D. Wellons, 1977, Wettability of dipterocarp
    veneers, Wood Sci. 10(1):22-27.

Jurecic, A., 1966, The basic principles of adhesion, Tappi
    49(7):306-310.

Kadlec, K. M., 1980. Wetting as a predictor of surface
    inactivation for platen dried Douglas-fir veneer, M.S.
    thesis, Oregon State University, Corvallis.

Kasper, J. B. and S. Chow, 1980, Determination of resin distribution
    in flakeboard using x-ray spectrometry, For. Prod. J., In
    press.

Kim, C. Y. and D.A.I. Goring, 1971, Corona induced bonding of
    synthetic polymers to wood. Pulp Pap. Mag. Can.
    72(11):93-96.

Lee, S. B. and P. Luner, 1972, The wetting and interfacial
    properties of lignin, Tappi 55(1):116-121.

Lehmann, W. F., 1968. Resin distribution in flakeboard shown by
    ultraviolet light photography, For. Prod. J. 18(10):32-34.

Luner, P. and M. Sandell, 1969, Wetting of cellulose and wood
    hemicelluloses. J. Polym. Sci. C 28:115-142.

Maloney, T. M., 1970, Resin distribution in layered
    particleboard. For. Prod. J. 20(1):43-52.
Marian, J. E., 1958, Adhesive and adhesion problems in par-
    ticleboard production, For. Prod. J. 8(6):172-176.
Marian, J. E., D. A. Stumbo, and C. W. Maxey, 1958, Surface
    texture of wood as related to glue-joint strength, For.
    Prod. J. 8(12):345-351.
Marian, J. E. and D. A. Stumbo, 1962a, Adhesion in wood.
    Part I: physical factors, Holzforschung 16(5):134-148.
Marian, J. E. and D. A. Stumbo, 1962b, Adhesion in wood.
    Part II: physico-chemical surface phenomena and the ther-
    modynamic approach to adhesion, Holzforschung 16(6):168-180.
Minemura, N., S. Hirata, S. Imura, and H. Takahashi, 1976,
    Studies on less formaldehyde liberation from Type II plywood
    glued with urea-formaldehyde adhesive, Rept. #65, Hokkaido
    Forest Products Research Institute.
Mizumachi, H. and H. Morita, 1975, Activation energy of the
    curing reaction of phenolic resin in the presence of woods.
    Wood Sci. 7(3):256-260.
Mizumachi, H. and M. Fujino, 1972, Interaction between wood
    and polymers, Holzforschung 26(5):164-169.
Mizumachi, H. and M. Kamidohzono, 1975, Dielectric properties
    of pva filled with wood components, Holzforschung
    29(6):229-231.
Moore, W. E., 1975, Problems in the determination of cured
    phenol-formaldehyde resins in paper and wood, For. Prod. J.
    25(4):39-41.
Nearn, W. T., 1974, Application of the ultrastructure concept
    in industrial wood products research, Wood Sci.
    6(3):285-293.
Nestler, F. H. M., 1977, The formaldehyde problem in
    wood-based products, an annotated bibliography, Report AD-A
    046077 US NT15.
Nguyen, T. and W. E. Johns, 1979, The effect of aging and
    extraction on the surface free energy of Douglas-fir and
    redwood, Wood Sci. Techn. 13:29-40.
Northcott, P. L., 1964. Specific gravity influences wood
    bond durability, Adhes. Age 1964 (October):34-36.
Northcott, P. L., H. G. M. Colbeck, W. V. Hancock, and K. C.
    Shen, 1959, Undercure-casehardening in plywood, For. Prod.
    J. 9(12):2-11.
Northcott, P. L., W. V. Hancock, and H. G. M. Colbeck, 1962,
    Water relations in phenolic plywood bonds, For. Prod. J.
    12(10):1-9.
Okuro, A., 1970, Adsorption of o-methylolphenol on wood surface.
    Bull. Govt. Forest Exp. Stn., Meguro, Japan 230:143-154.
Patton, T. C., 1970, A simplified review of adhesion theory
    based on surface energetics, Tappi 53(3):421-429.
Petersen, H., W. Reuther, W. Eisele, and O. Wittmann, 1974,
    Further investigations on the formaldehyde liberation during
    particleboard production with urea-formaldehyde adhesives.
    Holz Roh Werkst. 32:402-410.

Pillar, W. O., 1966, Determining curing properties of an
    adhesive in contact with wood, For. Prod. J. 16(6):29-37.

Plomley, K. F., W. E. Hillis, and K. Hirst, 1976, The
    influence of wood extractives on the glue-wood bond,
    Holzforschung 30(1):14-19.

Pohlman, A. A., 1974, Solid phase polymerization of ligno-
    cellulose and dibasic acids using acid activation, M.S.
    thesis, University of California, Berkeley.

Raffael, E., 1978, Progress in elimination of formaldehyde
    liberation from particleboard, Proc., Wash. State Univ.
    Particleboard Symposium 12:233-249.

Ramiah, M.,V., and G. E. Troughton, 1970, Thermal studies on
    formaldehyde glues and cellubiose-formaldehyde glue mixtures,
    Wood Sci. 3(2):120-125.

Rice, J. T. and R. H. Carey, 1978. Wood density and board
    composition effects on phenolic resin-bonded flakeboard,
    For. Prod. J. 28(4):21-27.

Salomon, G., 1965, Adhesion, in: "Adhesion and Adhesives,
    Volume I", Elsevier Publishing Co., Amsterdam, Netherlands.

Schwarz, F. E., R. L. Anderson, and A. G. Kageler, 1968,
    Resin distribution and how variations affect board quality.
    Proc., Second Wash. State Univ. Particleboard Symposium,
    Pullman.

Smith, L. A. and W. A. Côté. 1972. SEM and EDXA for the detec-
    tion of resin penetration into wood cell walls. J. Paint
    Techn. 44(564):71.

Schoring, P., E. Raffael, and G. Stegmann, 1972, Neuartige
    Holz-Zu-Holz-bindung in Holzwerdsoffen mit chemischen
    mehrstaff-systemen. Holz Roh Werst. 30(9):329-332.

Spalt, H. A., 1977, Chemical changes in wood associated with fiber-
    board manufacture, in: "Wood Technology: Chemical Aspects",
    I.S. Goldstein, ed., Symposium series #43, American Chemical
    Society, Washington, D. C.

Slaats, M., 1979, Eliminating glueline failure in bonding
    hardwoods. Adhes. Age 1979 (June):18-20.

Stamm, A. J., 1964, "Wood and Cellulose Science", Ronald
    Press, N.Y.

Stannett, V. T., 1967, Mechanisms of wet strength development
    in paper, in: "Surfaces and Coatings Related to Paper and
    Wood", R. H. Marchessault and C. Skaar, ed., Syracuse
    University Press, Syracuse, N.Y.

Steiner, P. R., 1973, Durability of urea-formaldehyde
    adhesives, For. Prod. J. 23(12):32-38.

Stofko, J., 1972, The autohesion of wood. Ph.D.
    dissertation, University of California, Berkeley.

Stumbo, D. A., 1964, Influence of surface aging prior to
    gluing on bond strength of Douglas-fir and redwood, For.
    Prod. J. 14(12):582-589.

Sundin, Birger., 1978., Formaldehyde emissions from par-
    ticleboard and other building materials, Proc., Wash. State.
    Univ. Particleboard Symposium. 12:251-273.

Tarkow, H. and C. Southerland, 1964, Interaction of wood
    with polymeric materials. For. Prod. J. 14(4):184-186.

Tarkow, H., W. C. Feist, and C. F. Southerland, 1966,
    Interaction of wood with polymeric materials--penetration
    versus molecular size, For. Prod. J. 16(10):61-65.

Troughton, G. E., 1969, Effect of degree of cure on the acid
    hydrolsis rates of formaldehyde glue-wood samples, J. Inst.
    Wood Sci. 23:51-56.

Troughton, G. E. and S. Chow, 1968, Evidence for covalent
    bonding between melamine formaldehyde glue and wood, J.
    Inst. Wood Sci. 21:29-34.

Troughton, G. E. and S. Chow, 1971, Migration of fatty acid
    to white spruce veneer surface during drying: relevance to
    theories of inactivation, Wood Sci. 3(3):129-133.

Venkateswaren, A., 1975, Adhesion of wood, Report OPX 122E
    of the Eastern Forest Products Laboratory, Ottawa, Canada.

Villaflor, A. A., 1973, Water loss in resin adhesives systems
    as related to wettability-permeability characteristics of
    wood substrates and to bond formation, Ph.D. dissertation,
    University of Washington, Seattle.

Weiner, J., 1977, "Wet Strength of Paper", Bibliographic series 168,
    Institute of Paper Chemistry, Appleton.

Wellons, J. D., 1977, Adhesion to wood substrate, in: "Wood
    Technology: Chemical Aspects", I.S. Goldstein, ed., Symposium
    Series 43, American Chemical Society, Washington, D.C.

Wellons, J. D., 1980, Wettability and gluability of Douglas-
    fir veneer, For. Prod. J. In press.

Wellons, J. D. and L. Gollob, 1980. GPC and light scattering
    of phenolic resins--problems in determining molecular
    weights, Wood Sci. In press.

Wellons, J. D., R. L. Krahmer, R. L. Raymond, and G. Sleet,
    1977, Durability of exterior siding plywood with Southeast
    Asia hardwood veneers, For. Prod. J. 27(2):38-44.

White, J. T., 1979, Growing dependency of wood products on
    adhesives and their chemicals, For. Prod. J. 29(11):14-20.

White, M. S., 1977, Influence of resin penetration on the
    fracture toughness of wood adhesive bonds, Wood Sci.
    10(1):6-14.

White, M. S., G. Ifju, and J. A. Johnson, 1977, Method of
    measuring resin penetration into wood, For. Prod. J.
    27(7):52-54.

Wilson, J. B., G. L. Jay, and R. L. Krahmer, 1979, Using
    resin properties to predict bond strength of oak
    particleboard, Adhes. Age 22(6):26-30.

Wilson, J. B. and M. D. Hill, 1978, Resin efficiency of com-
    mercial blenders for particleboard manufacture, For. Prod.
    J. 28(2):49-54.

Wilson, J. B. and R. L. Krahmer, 1976, Particleboard:
    microscopic observations of resin distribution and wood
    fracture, For. Prod. J. 26(11):42-45.

Young, R. A., 1978, Bonding of oxidized cellulose fibers and
     interaction with wet strength agents, Wood Fiber
     10(2):112-119.
Yutaka, I. and T. Yasuchi, 1977, Adsorption of polyethylene
     glycol on swollen wood. Mokuzai Gakkaishi 23(9):451-458.
Zisman, W. A., 1963, Influence of constitution on adhesion,
     Ind. Eng. Chem. 55(10):19-38.
Zisman, W. A., 1972, Surface energetics of wetting, spreading and
     adhesion, J. Paint Techn. 44(564):41-57.

CRITICAL PROPERTIES
AND THEIR DETERMINATION

PROBLEMS ENCOUNTERED WITH CONVENTIONAL

FIBER-REINFORCED COMPOSITES[‡]

Robert F. Landel
Jet Propulsion Laboratory
California Institute of Technology
Pasadena, CA   91103

The problems encountered in dealing with composites can be usefully classified as preparational, computational and operational. That is, we first ask how it is made and the effects of process variables, including the resultant chemistry, on the initial properties.[*] Then, knowing (or presuming) the component properties and attributes, one may try to calculate the properties of any given composite structure made from these and at the same time estimate its ability to withstand some generalized stress field, i.e., some combination of mechanical, thermal or environmental loads for some stated period of time. Finally, we can ask what problems are encountered under use conditions.

Preparational Problems

A. Resins

The thermosetting resins, especially epoxies and phenolics, are complex mixtures of chemicals, with initial functionalities of varying reactivity and with products of the reaction step able to

---

[*]Here, properties include moduli, thermal expansion coefficients, strengths and strength distributions, and their time dependence (both history of load/deformation and age since fabrication).

[‡]This paper represents one phase of research performed by the Jet Propulsion Laboratory, California Institute of Technology, sponsored by the National Aeronautics and Space Administration under Contract NAS7-100.

participate further in the reactions. Thus, an epoxy reacting with primary alcohol gives rise to a secondary alcohol which is potentially capable of further reaction with additional epoxy groups. Reaction to either the prepolymer state or to a B-stage can and usually does give rise to products which are a) not well understood chemically, and b) not reproducible. Thus "prepreggers," the manufacturers of the resin-impregnated tape used in composite lay-up, speak of the "state of advancement" at the time of final composite lay-up and cure. As might be expected, this varies with the time after initial preparation, even for materials stored in the cold. The use of high pressure gas-liquid chromatography is proving to be the most sensitive tool for separating and identifying initial and intermediate compounds and oligomers; such that, on the one hand, reactions can be studied, or, on the other, the HPLC trace can be used as a fingerprint to assure reproducibility or to isolate bad or overage batches.

Variations in the starting material can lead to difficulties in adequately coating the fiber tapes used in continuous fiber systems and to resin-rich or resin-poor or even voided regions in the final product. Changes in the cure condition can sometimes be made to alleviate such problems, but in some applications this is unacceptable, e.g., in the fabrication of every short-run high performance parts such as aircraft components.

As a corollary of matrix variability, it's propensity to absorb moisture and the ensuing effects of such moisture must be carefully considered. Chemists would not be too surprised to find that epoxies at the B stage can absorb 0.8% of water fairly quickly at ambient conditions and would expect that this water can affect the final product either (or both) by simply vaporizing during cure and creating voids or/and by changing the rate and course of the cure in a given cure cycle (leading to markedly different flow properties at different times in the cure cycle). Unfortunately, such wisdom did not permeate the industry until recently, following some unpleasant surprises.

Such changes in the flow behavior during cure, as evidenced by the viscosity of the resin, have been studied by measuring the dynamic viscosity vs. time for samples heated to a cure temperature or taken through a cure cycle. The viscosity initially drops, as the specimen heats to the cure temperature, then rises again when curing takes place. Molding has to take place during the interval when the viscosity is low, thus setting the time available for molding. On the other hand, the viscosity cannot be too low or the resin will flow out from between the fibers, leaving resin-starved, weak areas. The analysis of the flow behavior is quite complex, since one must deal with flow through a porous medium (D'Arcy flow) of a non-Newtonian liquid whose properties are steadily changing with time. In this sense, the problem has features reminiscent of reaction-injection molding (RIM).

In contrast to RIM and most other curing processes, where the reaction can go to completion or nearly so, the high T_g resin systems used with composites have a built-in reaction-stopper which can prevent complete reaction. That is, as the molecular weight builds up, the glass temperature and the viscosity of the resin system also increase. If the cure temperature is too low, i.e., below the T_g of the fully-cured matrix, then the reaction will stop when the T_g of the polymerizing system reaches the cure temperature. At this point the local mobility becomes too small to permit reactive sites to diffuse together. Even holding the specimen at the cure temperature will therefore not permit the cure to be completed. On cooling such incompletely cured specimens to room temperature, a high modulus, high strength product can result. However, the use temperature cannot exceed (or even approach) the cure temperature, otherwise it will once again soften and proceed to cure further. Such incompletely cured resins can lead to other problems, too, notably an enhanced moisture sensitivity.

The problem is not easily resolved by simply going to a higher cure temperature (or temperatures, in a step-wise cure schedule) because the fully cured material may actually be more brittle than the partially cured material and certainly the thermal contraction stresses which develop on cool-down will be increased.

B. Fibers

Taking glass, metal and graphite fibers in the "as-made" state, preparational steps and problems are concerned with preserving their initial properties and developing suitable surface pretreatments and bonding agents to provide an appropriate fiber-resin interface. Aramid fibers, being more reactive, offer a somewhat greater latitude in approaches to bonding agents, but their much lower transverse strength can lead to "overbonding" and fibrilar fracture in the fiber itself. (Similar problems could well occur with cellulosic fibers.)

An "appropriate" fiber-resin interface is a delicate matter. Assuming that the matrix properties away from the interface are not (or not significantly) altered by changing the bonding agent, in a system with high bond strength the fiber transfer length* is low, failure tends to initiate by fiber breakage followed by transverse cracking in the matrix and the quasi-static strength is high. Multiple transverse cracks are seen after failure and the failure

---

*When a unidirectional fiber system is loaded in tension in the fiber direction, the fibers carry the load; the load is transferred from fiber to matrix by shear at the interface. The broken end carries no load; the transfer length is the distance over which the load dwindles from it mean value to zero.

surface is relatively smooth. Conversely, with low bonding
strength, debonding occurs and cracks propagate along the fibers to
give a brush-like failure surface and a lower strength. However,
when the same systems are tested in fatigue, matrix cracking is a
more forgiving process and so longer fatigue lifetimes are
observed. Thus improved bonding can lead to higher strength at the
expense of reduced fatigue resistance.

As a corollary, low bond strength can lead to enhanced
moisture pickup at the interface, which on the one hand will weaken
the bond still further, yet on the other hand can plasticize and
toughen the matrix so that crack propagation can be reduced.

Computational Problems

There are, in fact, two regions to be considered: that of the
deformation prior to failure and of the failure region (Figure 1).
In the deformation region the properties of a material are given by
a description of the deformation. Here we distinguish between
mini-, micro- and macro-mechanics. Mini-mechanics refer to the
analysis of the stress in a single ply, i.e., in the fiber itself
and in the matrix between fibers (or particles). Micro-mechanics
refers to the analysis of the behavior of an assembly of fibers
-- a single ply. Macro-mechanics describes the behavior of an
assembly of plies, i.e., a laminate, or of a final structure.

On the other hand, when one reaches the failure region -- the
upper, dashed portions of the curve -- one is dealing with failure
mechanics in trying to describe the end-of-life and end use
conditions.

There are two points to be noted here: one is the fact that
for multiple tests under the same conditions, the results tend to
trace out a single stress-strain curve with the failure points
distributed along its upper end. Therefore, the failure stress or
failure strain is not a single value but a statistical distribution
of values. The actual value obtained for a given specimen is
simply one of the values from that distribution or population.

Secondly, in the region of incipient failure the deformation
can change the nature of the material itself and so cause a change
in the nature of the stress-strain response, i.e., different curves
are traced out. Both of these effects have to be accounted for as
part of failure mechanics.

It should be emphasized that in many cases we still do not
have good descriptions of the failure mechanics for
fiber-reinforced composites.

Now turning to the first region and considering the question
of characterizing the deformation, the underlying equations and
relationships -- the underlying theory -- for calculating the
mechanical response is rather well understood. In particular, the

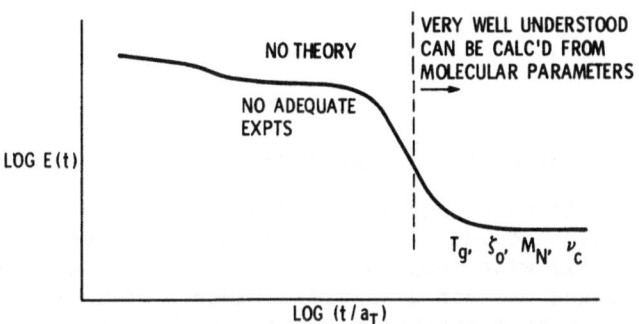

Figure 1. Deformation and failure regions--two aspects of the problem of describing mechanical properties.

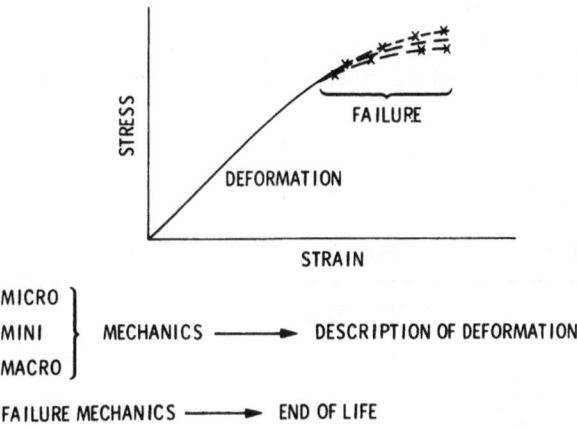

Figure 2. Sketch of the time dependence of the modulus, which can drop by two or three orders of magnitude in the principal transition region. The minor transition in the glassy region is often very small or missing entirely.

underlying theory for calculating the mechanical behavior of
laminae from the moduli of the components and that of the laminate
from the properties of the laminae and their orientation appears to
be reasonably well in hand.[1] Functionally, the relationship
between the stress σ and strain ε tensors is given by:

$$\sigma_{ij} = C_{ijkl}\varepsilon_{kl} \tag{1}$$

where

$$C_{ijkl} = C(E_1, E_2, \nu_1, \nu_2, \theta_n, S, \ldots) \tag{2}$$

That is, the stiffness matrix C is dependent on the matrix modulus
E_1 and the fiber modulus E_2 and their Poisson's ratios, the angles
θ_n of the nth ply with respect to some reference angle and the
stacking sequence S. For a specific example of these function
relationships in an orthotropic plane stress situation, eqn. 1 can
be written as follows:

$$\begin{Bmatrix} \sigma_1 \\ \sigma_2 \\ \tau_{12} \end{Bmatrix} = \begin{bmatrix} Q_{11} & Q_{12} & 0 \\ Q_{12} & Q_{22} & 0 \\ 0 & 0 & Q_{64} \end{bmatrix} \begin{Bmatrix} \varepsilon_1 \\ \varepsilon_2 \\ \gamma_{12} \end{Bmatrix} \tag{3}$$

where the individual components Q_{ij} are related to the ordinary
elastic quantities, moduli and Poisson ratios, by the following:

$$Q_{11} = \frac{E_1}{1 - \nu_{12}\nu_{21}} \qquad\qquad Q_{66} = \varepsilon_{12}$$

$$Q_{12} = \frac{\nu_{12}E_2}{1 - \nu_{12}\nu_{21}} = \frac{\nu_{21}E_1}{1 - \nu_{12}\nu_{21}} \qquad \nu_{12} = \frac{-\varepsilon_2}{\varepsilon_1} \tag{4}$$

$$Q_{22} = \frac{E_2}{1 - \nu_{12}\nu_{21}}$$

Next we must concern ourselves with the parameters of the
stiffness matrix and determine whether they are constant. In fact,
they are not. Thus, if we concern ourselves only with the modulus
of the matrix itself, then we have the response indicated in Figure
2. Here the modulus is sketched as a function of an effective time
scale called the reduced time scale, where the parameter $1/a_T$ is a
"shift factor" which changes the time scale according to the
temperature of the experiment. The modulus at very short times or
very low temperatures is high, the material being glass-hard and
relatively tough. With increasing time I have sketched for
generality a transition such that at sufficiently long times or
high temperatures the modulus drops by 2 or 3 orders of magnitude

to a rubbery modulus characteristic of a crosslinked system. The time-temperature superposition, i.e., the shift factor a_T, applies rather well from roughly the midpoint of the transition zone on out. Furthermore, for this region the properties are very well understood. There are well-developed theories for the behavior and, in fact, in some cases the properties can be calculated from just a few molecular parameters such as glass temperature·T_g, monomeric friction factor, molecular weight between entanglements (all of which are specific to a given polymer) and the crosslink density (which is characteristic of a given preparation). In the glassy zone, however, there are two major deficiencies: (1) there is no molecular theory and (2) there are no adequate experiments. By the latter, I mean that the usual experiment is done at a single frequency or over a very limited time scale, and the principle experimental variable is the temperature. What is needed are isothermal experiments over very wide ranges in time scale in order to separate the time and temperature behavior.

In general, in the glassy region, the modulus is roughly linear in log time, i.e.,

$$E(t) = \log E_O - m \log t \tag{5}$$

The time dependence of compliance, the reciprocal of the modulus, can be expressed in a similar form of equation. It turns out that the compliance is more useful than the modulus in calculating the mechanical properties; in fact, converting from the compliance back to the modulus, in order to express the response in terms of moduli, can lead to complications in the alegbra of the resulting representation.

Two points should be made in regard to the behavior indicated in Figure 2. The first is that, in addition to the temperature dependence there is a similar dependence of the modulus of the unfilled matrix on moisture content. Thus an increase in moisture can plasticize the material, and plasticization can shift the total curve to the left and at the same time may decrease the level of the rubbery plateau modulus. In fibrous composites the absorption of water can lead not only to changes in the material properties, but also to changes in the stress states as well. That is, as indicated in Figure 3, when heat or moisture is applied, an individual ply will expand slightly in the longitudinal direction and a great deal more in the transverse direction.[1] When the plies are laid up in a cross-ply arrangement, the resultant deformation is a compromise between the two so that the resin in both plies tends to be in compression. These thermal-and moisture-originating expansive and contractive stresses must be accounted for in describing both the overall load on the final piece or structure and the localload on the boundaries between plies or the boundaries between matrix and fiber reinforcement. Figure 4 illustrates the anisotropic dimensional changes which can occur, even on a cross-plied laminate, when exposed to moisture.

An underlying question here is that of the rate of permeation into and out of the composite -- when is it Fickian, when

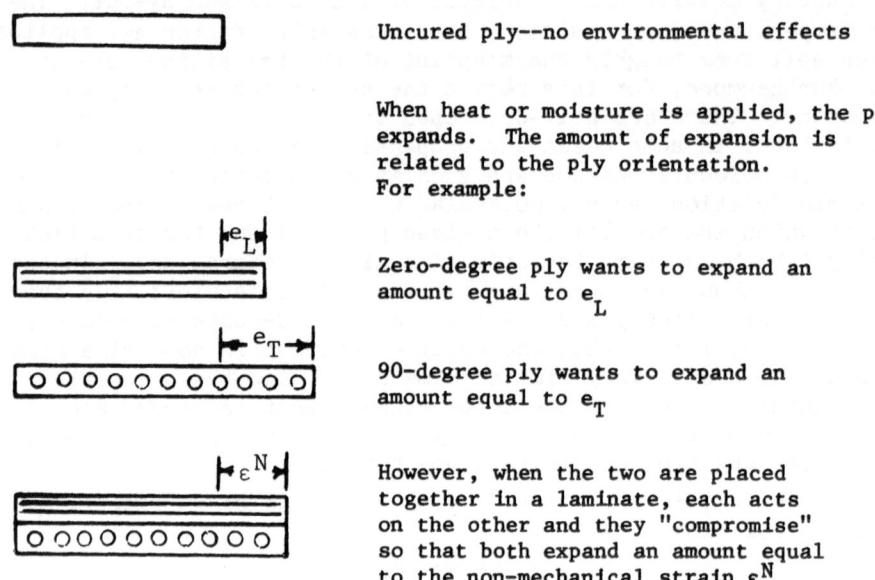

Uncured ply--no environmental effects

When heat or moisture is applied, the ply
expands. The amount of expansion is
related to the ply orientation.
For example:

Zero-degree ply wants to expand an
amount equal to e_L

90-degree ply wants to expand an
amount equal to e_T

However, when the two are placed
together in a laminate, each acts
on the other and they "compromise"
so that both expand an amount equal
to the non-mechanical strain ε^N

Figure 3.   Non-mechanical strains in a laminate.

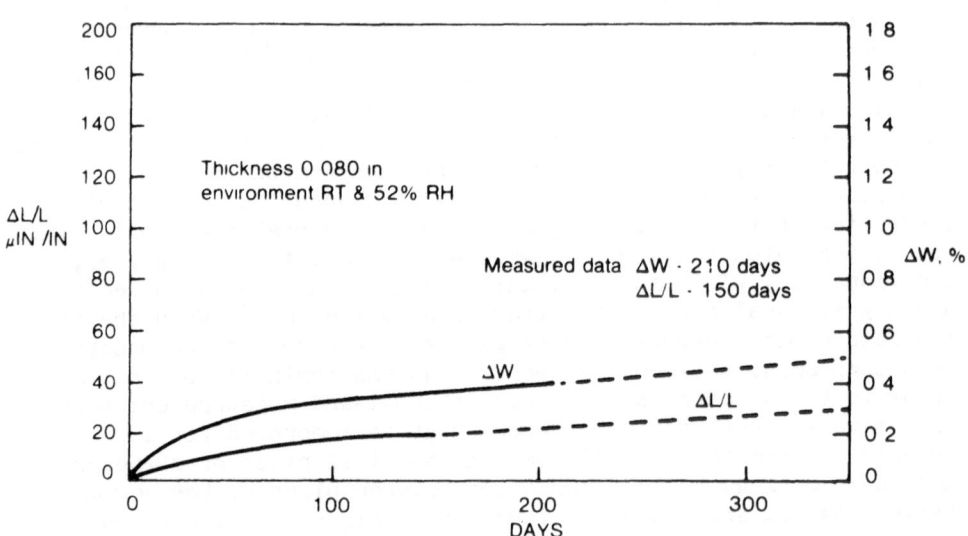

Figure 4.   Environmental effects on $[(0/45/90/135)s_2$ GY-70/X-30.

non-Fickian, and why? For example, in a strained specimen
penetration can be faster than that in the unstrained state because
of the presence of microcracks; but simply cycling moisture into
and out of the specimen withou engendering cracking is also known
to change the permeation rate, presumably because of a change in
local molecular ordering.

A probably even more important process for modifying the
modulus, which is just now beginning to be studied (and mostly in
thermplastics, not thermosets), is that of physical ageing. Thus,
as indicated in Figure 5, when a material is cooled from the liquid
state to a temperature below T_g, the material goes into the glassy
state. However, in this state it can continue to contract toward
the equilibrium, liquidus line. The extent of the initial
deviation from the equilibrium behavior will depend on whether or
not the sample was quenched from the liquid state or slow-cooled.
In any event it will not be in equilibrium but will be attempting
to diminish its volume toward this state. Thus after specimens or
structures are removed from cure and cooled down, there is a steady
unstoppable volume decrease in the resin (to points 1, 2 and 3 of
the Figure). The rate of decrease depends on the rate of cool-down
from cure and all subsequent thermal (and moisture) history, though
it is a steadily decreasing function of time at any fixed
temperature or water content. Practically, this means on the one
hand that impact resistance and strength are higher immediately
after fabrication -- e.g., this is the best time to cut bolt holes
or (in short fiber systems) to do post-forming. Operationally, it
turns out that the volume diminution affects the creep behavior.
As shown in Figure 5, the initial creep is higher and passes from
secondary to tertiary creep sooner, the sooner the sample is tested
after cool-down. Theoretically, this volume shrinkage means that
E_2 and ν_2 in eqn. 1 (which really depend on shear and bulk moduli
G_2 and B_2, or the shear modulus and the Lame constant) are,
additionally history dependent. Hence, ageing time A must be
considered in eqn. 2 along with the well known mechanical
relaxation time τ and the (usually not considered) chemical ageing
time A_c. The latter will be important for situations in which
further chemical reaction can take place in the system, e.g.:
extremely long time exposure under normal conditions (oxidative),
exposure to chemical attack, conditions under which post-cure
reactions can take place, and under gamma or uv radiation.

Failure

A. Mechanisms

Serious difficulties arise when one tries to account for resin
cracking, debonding and fiber fracture. These local failures lead
to a reduction in the effective modulus and a deviation from linear
response which can become quite large. A factor of two is not
uncommon. The physical origin of these irreversible processes, of
their propagation and termination and their dependence on localized

non-filled and each for a series of a at stress and temperatures (?). Here can be seen in the comparison of the
so the measured to the mixture and resultingly a tower and
and one of the some constituent's representing the slope by the effect
so change the particular that is (relatively represented a constitution to the set. ...

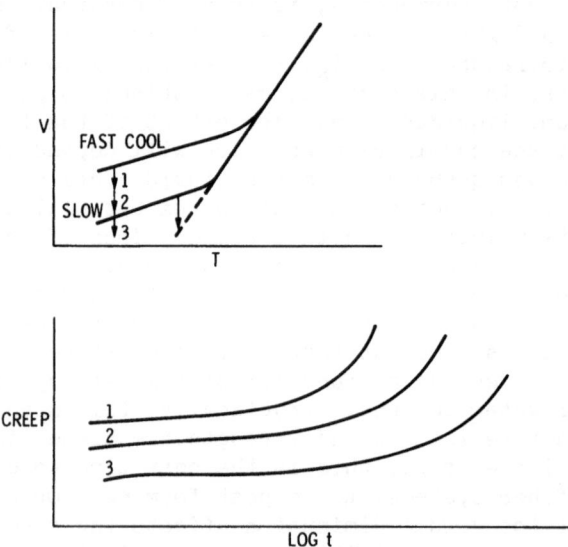

Figure 5.   Volume-temperature relations, creep and physical
            aging.

stress fields and environments, are frequently ill-understood and poorly modeled. Not only do non-linear responses result, but unloading and reloading behavior can be drastically different and the extent of the devation from linearity will depend on the number of work-in cycles (i.e., the number of cycles to some quasi-study state). Moreover, these effects will, of course, depend on the composition and construction of the part or structure under consideration. When the part must withstand repeated loads into this region of damaging load or deformation, it is the fatigue behavior which is important.

Mechanistic models for fatigue behavior are in a rather primitive state; mathematical-phenomenological models are in a considerably better state. Indeed, the industry may be suffering from a plethora of incomplete models!

B. Description

Failure, as indicated in Figure 1, represents a limiting point on the curve or a limiting stress value. Since, as we saw in eqn. 3 for the plane stress situation, there are three separate values of the stress, there are therefore three separate values of the failure stress.

These three values are interrelated, just as the stress-strain response is dependent on the contributions of stress in the three directions. Hence a plot of the breaking stress is a three dimensional geometric figure which represents a limiting bound to the failure stresses. There are many such descriptions of failure surfaces, e.g., von Mises, maximum principle stress, maximum principle strain, etc. One indicative failure description is that of Tsai and Hill.[2] For a plate in plane stress, made of unidirectional lamanae, the Tsai/Hill failure criterion is given by:

$$\left(\frac{\sigma_1}{\sigma_{1b}}\right)^2 + \left(\frac{\sigma_2}{\sigma_{2b}}\right)^2 - \left(\frac{\sigma_1\,\sigma_2}{(\sigma_{1b})^2}\right) + \left(\frac{\tau_{12}}{\tau_{12b}}\right)^2 \qquad (6)$$

where τ is the shear stress, the subscript b denotes the breaking stress for a particular direction and the unsubscripted values denote the corresponding applied stress. This may be expanded, in the case where the load is off-axis to the principle fiber direction by the angle θ, to the following expression:

$$\frac{\cos^4\theta}{\sigma_{1b}^2} + \left(\frac{1}{\tau_{12b}^2} - \frac{1}{\sigma_{1b}^2}\right)\cos^2\theta\,\sin^2\theta + \frac{\sin^4\theta}{\sigma_{2b}^2} = \frac{1}{\sigma_x^2} \qquad (7)$$

These equations describe a cigar-shaped surface whose major axis
lies roughly along the σ_1 direction. Such a surface for a
glass-epoxy system is sketched in Figure 6.[3] One can then apply
any combination of stresses and not suffer failure in the plate so
long as that combination does not exceed the combination indicated
by this boundary.

The question then arises as to how the failure surface will
vary with time or with moisture, etc., because it is apparent that
if the moduli (the coefficients in eqn. 3) vary under these changed
conditions then the failure surface must also change. An
illustration of such a change is indicated in Figures 7 and 8 where
the fiber-glass specimens were soaked in benzene to plasticize and
weaken the material. Figure 7 shows that for the 0` orientation
and for a relative loading of 80% of the initial value, benzene has
reduced the failure stress level or, equivalently has caused the
material to break sooner. At a slightly higher relative loading
value, the effect has been rather small. This decrease in failure
stress values translates into a shrinkage of the failure surface.
Figure 8 shows how the surface contracted with increasing time
under load.

C. Fracture Mechanics

Fracture mechanics is concerned with the propagation of cracks
-- pre-existing cracks or cracks which can develop from a flaw and
its effect on the strength and time dependence of strength. The
general analysis (see Figure 9) considers the crack, and the
damaged region both at the root of the crack and ahead of it,
extending over some distance d. The energy required to form a new
surface Γ will be dependent upon, it turns out, the creep
compliance, $D(t_c)$, where the time scale t_c now refers to time
required for the crack to propagate the distance α, and a stress
intensity factor K_I^2 which is a measure of the amount by which the
local stress is increased over the boundary stress. Thus

$$\Gamma \propto D(t_c)\, K_I^2 \qquad\qquad (8)$$

From considerations such as these and the incorporation of the time
dependent response of the material, one can calculate quantities or
results such as the crack velocity as a function of initial crack
length or subsequent crack length, or the stress required to cause
a crack to grow to failure. In one such formation, due to
Schapery,[4] the crack velocity $\dot{a}$ is a function of the creep
compliance D and the stress intensity factor K:

$$\dot{a} = \dot{a}\left(\frac{D_2}{D_0}\right)^{1/m}\left(\frac{K_I}{K_{I_0}}\right)^{2(1/m)}, \qquad\qquad (9)$$

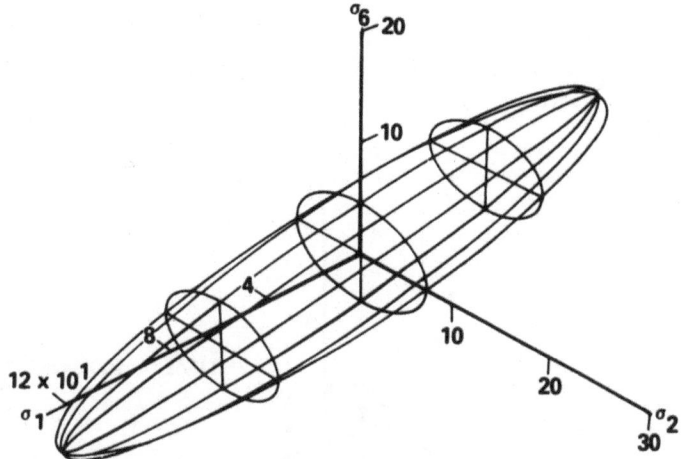

Figure 6.   Quasi-static failure surface for glass–epoxy
composite material (F_{12} = 1/2 stability limit).
($\sigma_6 = \tau_{12}$)

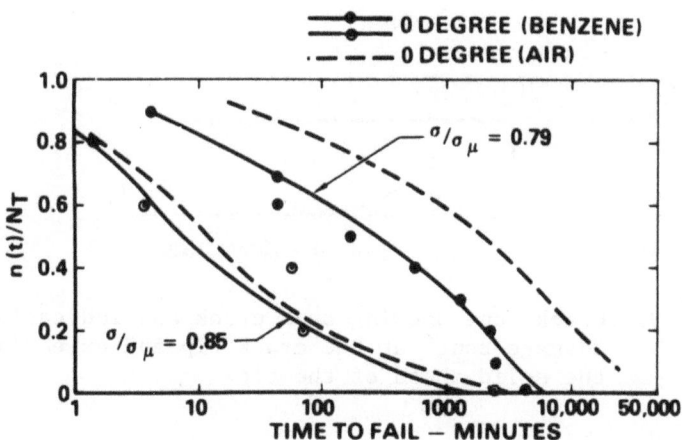

Figure 7.   Cumulative distribution in time to failure for
0-deg specimens tested in benzene environment.
Air environment tests shown for comparison.

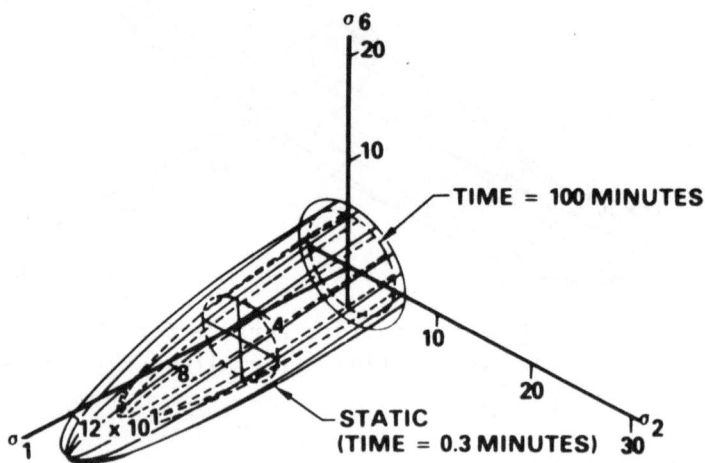

Figure 8.   Effect of time on the first-quadrant failure surface.

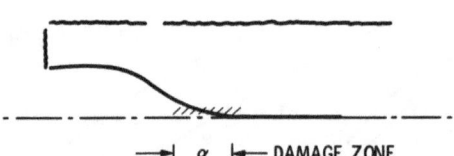

Figure 9.   Crack zone showing open crack and indicating the
            "damage zone" at the crack tip and extending into
            the solid ahead of the tip.

where D_O is the initial compliance of the material and K_I is the stress intensity factor based on it. Of course as the crack propagates farther and farther through the sample the stress intensity factor rises accordingly. These results can be translated into calculations of the time that the sample would withstand a stress of a given value. Thus as indicated in Figure 10, the stress which can be safely applied is a decreasing function of the exposure time.

Similar expressions can be developed to calculate the effects of cyclic loading. Thus if one assumes for simplicity that the crack velocity is proportional to K_I^q, and if K_I^O is independent of the stress level, Schapery has shown that an expression equal to Miner's law follows directly.

If we consider the creep failure curve of Figure 10, the question arises as to what happens if the stress is applied for a period of time less than the failure time, then is reduced and held at that level for another period of time and then reduced still further for the third period. At the end of these periods, one would like to know the lifetime remaining if a fourth stress were applied. According to Miner's law, one would calculate the fraction n/n_f for each of the steps where n is the duration of time the stress is applied and n_f is the life time under that load if only it had been acting. The sum of these fractions would be equal to unity, allowing us therefore to calculate the residual life under step four. In fact however, Miner's law rarely holds and must be modified to account for the type of load (e.g., tension only or mixed tension-compression) as well as load sequence (i.e., stress 2 > stress 1 or the reverse). In Schapery's formulation the load sequence is accounted. Neither Schapery's nor equivalent theories have been subjected to sufficient tests to determine their ultimate validity or utility, but they certainly offer a very good and very rational starting point.

Unfortunately, in all of the above, the mathematical or arithmetic computational complexities are such that it is not easy to apply these theories in a direct form. What is required is either simple ways of applying them or pre-computed engineering design curves from which intervening values can be estimated. In other words, practical means of utilization of these theories and techniques must be developed. A very significant step in this direction has recently been taken by Tsai and Hahn of the Air Force Materials Laboratory, who have developed simple coding schemes which will allow a variety of stress analysis problems in cross-plied laminates to be evaluated on a preprogrammed hand-held calculator.

The underlying equations can be recorded on magnetic tape (i.e., for an HP67 or a T159). Hence the engineer now has at his fingertips a technique which takes the drudgery out of the task so that the desk top calculations can be made to gain initial

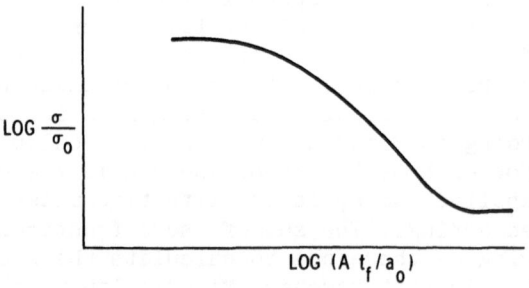

Figure 10.   Sketch of the life time of a sample under a given
             load.

estimates of properties and responses. This approach is strongly
to be encouraged and recommended and hopefully future developments
will quickly appear to expand the direct range of applicability of
the hand-held calculator and the microcomputer.

Operational Problems

Operational problems center on the detection, evaluation and
repair of flaws. The latter can arise normally, in the sense that
one expects cracks to develop during operation even when the
initial design is sound, or abnormally. In the latter case, there
is no way to design around the problem -- examples of this include
the dropping of tools onto components or the effect of the use of
the wrong solvents for cleaning. In such cases detection is not a
problem, but this is not true in the former case, where flaws or
cracks can develop in inaccessible places such as around bolt holes
but underneath the bolt head and washer or in submerged ends of
panel pieces. Even in accessible places, some types of flaws are
readily detected while others are detected with great difficulty.
What is needed then, are means both of detecting critical flaws,
and of deciding which flaws are critical. In other words,
evaluating the seriousness of flaws once they have been discovered
and their extent has been delineated.

As for repair of damage, we do not have good repair techniques
for fiber reinforced composites. In simple cases, e.g., an impact
damage area, it may be possible to cut out the damaged area and
install a patch. In other cases a patch "doubler" can be installed
over the damaged area. However, it is not clear how one should
best handle delamination damage. These situations are perhaps
analogous to repairing wood structures -- while it is not clear
that a badly checked beam end could not be repaired with an
adhesive, currently the damaged end is excised and replaced. Were
the damage to occur interior to a structure, in a load-bearing
section, repair would be much more complicated.

Summary

In order to overcome problems which arise in the preparation
of composites, more information is needed on the reactions that are
taking place and the effects of such reactions on modifying both
the network structure and the morphology of the matrix.

Computational problems arise because: a) not enough is known
of matrix failure mechanisms or of the influence of molecular
structure/topology on the coefficient used in the calculations and
b) theories for interlaminar failure are incomplete.

Operationally, there is a need for better means of detecting
and intepreting flaws and for useful means of repairing damage.

References

1.   For example, Jones, R. M., "Mechanics of Composite Materials,"
      McGraw-Hill, New York, 1975 and Christensen, R. M., same
      title, Wiley-Interscience, New York, 1979.

2.   Tsai, S. W. and Hahn, H. T., "Introduction to Composite
      Materials," AFML-TR-78-201, Air Force Materials Lab,
      Wright-Patterson Air Force Base, Dayton, OH.

3.   Wu, E. M. and Ruhman, D. C., "Stress Rupture of Glass-Epoxy
      Composites: Environmental and Stress Effects," in "Composite
      Reliability," ASTM STP580, ASTM, Philadelphia, PA, p. 263.

4.   Schapery, R. W., "Deformation and Failure Analysis of
      Viscoelastic Composite Materials," _From an Elastic Behavior of_
      Composite Materials, ASME Publication, AMD-Vol. 13, Chapter 5,
      ASME, New York, New York (1975).

5.   Tsai, S. W. and Hahn, H. T., "T1 59 Magnetic Card Calculator
      Solution to Composite Materials," March 1979, Air Force
      Materials Lab, Wright-Patterson Air force Base, Dayton, OH.

WOOD COMPOSITES

Robert H. Gillespie

Forest Products Laboratory
P.O. Box 5130
Madison, WI 53705

INTRODUCTION

Wood composites consist of various kinds of wood elements bonded
together in different combinations and configurations. The elements
may be fibers, particles, flakes, wafers, strands, veneers, or sawn
lumber from a wood resource. These elements may be combined with
metal foils, plastic films or foams, or a variety of other materials
to achieve some specific performance requirement. The different
wood elements can also be combined in various ways to provide the
properties desired for a particular end use.

Wood composites are developed to fill a need usually associated
with long-term applications, such as those required by building components. Consequently, they take the form of sheets or panels to
serve as covering material, or they take the form of dimension lumber
or timbers to provide framing and structural elements.

Typical sheet materials are particleboards, flakeboards, wafer-
boards, plywood, or composites having random particles or oriented-
strand cores with veneer faces. Flakeboards may have randomly
distributed flakes, oriented flakes, or different flake sizes and
orientations in the faces than in the interior of the sheet. Structural elements may consist of veneers or lumber laminated in parallel
grain configurations or made up of thick particleboard with veneered
edges. Other structural components consist of combinations such as
I-beams, box beams, or stressed-skin panels formed by bonding thin
faces such as plywood or particleboards to framing members or as
sandwich panels having thin faces bonded to low-density core
materials such as plastic foams or resin-treated paper honeycombs.

The opportunities are innumerable to design special properties into new wood composites for subsequent assembly into unique building components.

Wood composites have been developed in the past for both interior and exterior applications with the materials for interior use commonly bonded with urea-formaldehyde resin adhesives or casein, and those for exterior use bonded with phenolic-type adhesives. The increasing use of phenolic-type adhesives in recent years reflects an emphasis on products that will provide long-term performance in severe service environments.

Although urea-formaldehyde, casein, and phenolic-type adhesives have been the principal bonding agents of choice for wood composites in the past, many other types will play a role in the future. The conventional adhesives are characterized as thermosetting materials of high strength and high rigidity requiring thin bondlines for the development of maximum properties. But thin bondlines cannot always be achieved, particularly when wood-based materials are assembled without opportunities to prepare well-fitted surfaces. A case in point is the development of diaphragm-type floor systems involving the bonding of plywood panels edge to edge to form a continuous diaphragm and nail-bonding of this diaphragm to the supporting joist. This is a wood composite having greater strength and stiffness than similar construction assembled with mechanical fasteners alone. The bonding is accomplished with gap-filling elastomeric-type construction adhesives whose strength and rigidity are less than that of wood. Thus there are potential applications for adhesives with widely varying properties in either primary bonding of a wood composite or in secondary bonding where wood elements are being assembled in a building component.

The properties and performance of a wood composite therefore depend upon the properties of the bonding material and particularly on whether or not the adhesive is higher or lower in strength and stiffness than the wood being bonded. In predicting the performance of a wood composite, the design engineer ignores adhesive properties when they are stronger and stiffer than those of wood. The adhesive serves only as a stress transfer element, and the properties of the adherend govern the composite's performance. But with adhesives less strong and stiff than wood, their properties must also be taken into consideration in predicting the performance of a bonded composite. Consequently, there is a need to consider not only the critical properties of wood components but also, in some cases, the critical properties of the bonding agents themselves as a prelude to predicting wood composite performance.

The critical properties required of wood composites used in severe service environments are many and varied. Strength properties are usually paramount and include rupture strength in bending, shear, tension, and compression. For some applications, stiffness or rigidity is a prime requisite. Internal bond strength, impact strength, and fastener-holding ability may be important in some applications, while racking strength, fire performance, acoustic, and insulating properties may be required in others. Creep properties and load duration limits are critical in some applications.

Of utmost importance in most applications are the durability properties, including resistance to the chemical effects of aging as well as the physical effects of loads imposed on joints by either internal or external forces. Incidental to the requirements for strength and performance are a number of other requirements such as density, size, squareness, appearance, moisture content, moisture absorptivity, and dimensional stability.

Properties critical to the performance of wood composites may be categorized in different ways. But for the purposes of this discussion, they will be classified as either physical and appearance properties, strength and stiffness properties, effects of moisture on dimensional stability, performance properties, and durability considerations. While the property measurements need to vary in detail from one type of product to another, those described for panel products such as particleboard and plywood will be emphasized. Also, test procedures from sources in the United States will be referenced in large part because of the author's awareness of U.S. sources. Similar test procedures are employed in other countries but it is not within the scope of this paper to include a discussion of the detailed differences that may exist.

PHYSICAL AND APPEARANCE PROPERTIES

All specifications and product standards contain requirements about size, squareness, thickness, surface appearance, and allowable variations in these characteristics to describe products suitable for the marketplace. These properties are critical for performance insofar as they assist in the judicious selection of materials and their correct installation with the view toward preventing misuse and improper application of products. Such specifications and standards also contain information relative to the intended use of the product described, the various types available, markings for identification, and often directives for application.

Other properties essential to proper selection and use include specific gravity, or density, and moisture content. Small representative specimens of the wood composite, which are accurately weighed and dimensions accurately measured, are again weighed after oven-drying in an oven at 103 $\pm$ 2° C until approximately constant weight is attained. The moisture content is then calculated on an ovendry weight basis and specific gravity based on volume at test and weight when ovendry.[1]

STRENGTH AND STIFFNESS PROPERTIES

In many of the strength and stiffness measurements, directional properties of boards can be determined by cutting specimens from appropriate orientations. Measurements are often made on both wet and dry specimens.

Static Bending[1]

In static bending tests the specimens are loaded at a slow specified rate by center loading over a span 24 times the nominal thickness of the specimen. A complete load-deflection curve is obtained until specimen failure. From the data obtained, and knowing specimen dimensions, calculations can be made for stiffness or modulus of elasticity (MOE), strength or modulus of rupture (MOR), stress at proportional limit, and work to maximum load.

The flexural properties of plywood[10] may be determined by any of three methods. A center-point flexure test for small simply supported specimens provides total deflection and modulus of elasticity data containing a relatively constant component attributable to shear deformation. A two-point flexure test yields deflection and modulus of elasticity data related to flexural stress only without a shear component. A third variation is a pure moment test for large specimens which is also free of shear deformation effects. The measured properties are of importance in most structural uses such as construction of floors, wall sheathing, roof decking, concrete forms, space and plane structures, stress-skin panels, and containers, crates, or pallets.

Tensile Strength Parallel to the Surface[1]

This measurement requires specimens having a reduced cross-section midway between the grips of the test machine, sometimes referred to as dumbbell shape. The specimens are loaded in tension at a low specified rate until failure.

Tests have also been standardized for plywood[14] using similar types of specimens with reduced cross-sections in the center of the long dimension. An alternate procedure involves large panel tests to evaluate manufacturing variables and growth characteristics that influence tensile properties.

Tensile Strength Perpendicular to the Surface[1]

This measurement, which is often called an internal bond (IB) strength test, is a measure of the cohesive strength of a board perpendicular to its plane. Loading blocks of steel or aluminum alloy are bonded with a suitable adhesive to both faces of a square specimen. Special fixtures attached to the heads of a testing machine engage the loading blocks and the specimens are stressed in tension to failure.

Compression Strength Parallel to the Surface[1]

Different procedures can be selected depending on the thickness of material to be evaluated. Thin materials may be tested with lateral support or laminated to provide a nominal thickness of at least 1 inch. The length of the specimen or height as loaded should be four times the nominal thickness. Width and thickness of the specimen are accurately measured to calculate the net cross sectional area. Specimens are loaded in compression to failure with the aid of a self-aligning spherical loading block. From the data obtained, modulus of elasticity, stress at proportional limit, and stress at maximum load may be calculated.

The compressive strength of plywood[3] in response to stresses acting in the plane of the surface involves both small-specimen and large-specimen testing. Both maximum crushing strength and elastic properties may be determined. The small specimens are supported laterally, without exerting undue pressure against the sides of the specimens, to prevent buckling during the collection of load-deformation data. The compressive properties obtained with full-size panels provide strength data that can be used for design purposes.

Shear Strength in the Plane of the Board[1]

The intent of this test is to have the shear plane parallel to the surfaces of the board with failure, after compression loading, taking place midway between the two surfaces. Shear blocks with a small square shearing area are prepared by lamination. The blocks are notched and placed between the anvils of a special jig adjusted to promote failure along the desired plane. In a related test for laminated products where bondline shear strength is required, similar block-type shear specimens are employed with notching and anvil placement adjusted to promote failure within the bondline.[4]

A somewhat more sophisticated approach to the measurement of shear strength in the plane of the board involves the bonding of steel loading plates to the two board surfaces. A length-to-thickness ratio of 12:1 is prescribed as a minimum so that secondary normal stresses are minimal when the specimens are loaded in

compression in a special jig. The loading system and jig arrangement
provides a plane through which the load acts extending from one edge
of the specimen to the diagonally opposite edge. The load-deformation
data obtained allows the calculation of an interlaminar shearing
modulus, a secant modulus or other type, and the interlaminar shear
strength. A similar test specifically for plywood to evaluate rolling
shear or shear in the plane of plies has also been standardized.[9]

Edgewise Shear Normal to the Plane of the Board[1]

This test produces the kind of shear stressing that occurs when
shear forces are introduced along the edges of a material. Two pairs
of rigid metal rails are clamped to specimens and compression-loaded
to introduce shear forces along the edges of a specimen. This simu-
lates applications where racking forces are involved. A similar test
has been standardized for measuring shear through the thickness of
plywood.[7] Alternative methods include a panel shear test for small
specimens and a panel shear test for large specimens. The strength
and modulus of rigidity values so obtained are useful for the rigorous
design of lumber-plywood components such as trusses with plywood
gussets, I-beams, box beams, folded plate roofs, space and plane
structures, floor and roof diaphragms, and shear walls. These
properties are strongly influenced by the depth of knife checks
resulting from the veneer cutting process.

Hardness and Hardness Modulus[1]

Hardness properties are measured by a modified "Janka ball"
test in which a ball is loaded to penetrate the surface of the
material under evaluation. Thin materials require lamination to
provide sufficient thickness to test. Autographic recording of
load-penetration data allows the calculation of (1) hardness modulus
as the load required per unit depth of penetration and (2) hardness
as the maximum load to embed the "ball" to one-half of its diameter.

Impact Resistance

The falling ball impact test[1] measures the resistance of boards
to the kind of damage that occurs in service when boards are struck
by moving objects. A steel ball of specified diameter is dropped
onto a supported panel or board from increasing heights, dropping
on the same center point of the panel until failure occurs. The
height of drop that produces a visible failure on the opposite face
to that receiving the impact is an index of impact resistance.

There is another method for measuring the toughness of plywood.[8]
A plywood beam specimen is simply supported and loaded at midspan,
and the energy necessary to cause failure by impact loading is
measured. The load is supplied by a pendulum so arranged that a
measured amount of the energy from its fall impacts the specimen.

A dropping sandbag impact test provides for the evaluation of entire wall, roof, or floor assemblies including framing members and coverings.[9]

Fracture Toughness

Until recently, the mechanical properties of materials were studied in terms of stresses and strains, strength and stiffness, or forces and distances. There are many cogent arguments to suggest that material properties should be studied in terms of energy, using a fracture mechanics approach.[11] Using this concept, materials may be tested in one of their weakest failure modes, cleavage, and critical strain energy release rates determined. Fracture toughness, crack initiation, and crack arrest values may be compared to provide a measurement of brittleness. This approach has been applied to the evaluation of oriented flakeboard by Lei and Wilson,[12] and to wood-adhesive bonds by White and Green.[13] A uniquely useful specimen was developed for fracture tests and used to study the effects of bonding variables on the fracture energy of phenolic-bonded wood joints by Ebewele, River, and Koutsky.[14]

Long-Term Load Deformation

Methods have not been standardized for determining the long-term load properties of wood composites providing creep, fatigue, and duration of load data under continuous or cyclic load situations. Since these important properties of materials will be covered in detail by R. G. Pearson, reference will be made only to a few examples of reported work. McNatt studied the effect of rate of loading and duration of load properties of particleboard,[15] Kufner reported on creep in particleboard,[16] Pearson investigated duration of load on the bending strength of wood,[17] with more recent data reported by Gerhards on both wood and wood-based materials.[18]

FASTENER HOLDING PROPERTIES

Fastener holding tests specify a particular size and type of nail or screw. However, they are adaptable to evaluation of other sizes and types of mechanical fasteners with suitable modification of procedures. In most cases, specimens are evaluated in both the dry and water-soaked conditions.

Lateral Nail Resistance[1]

This test measures the resistance of a material to the lateral movement of a nail that has been driven at right angles to the face of the board. The nail is driven through the board to extend an equal distance from each face so the nail can be engaged by a stirrup attached to the head of a testing machine. Specimens are prepared with the nail at three different and prescribed distances from the edge. The load required to move the nail to the edge is a measure of the resistance to lateral movement.

Nail Withdrawal Test[1]

This test measures the resistance to withdrawal of a nail driven through a board from face to face normal to the plane of the board. The nail is driven only to a depth such that the head of the nail can be engaged by a load-applying fixture attached to the head of a testing machine. The maximum load to withdraw the nail at a uniform rate of loading is determined.

Nailhead Pull-Through Test[1]

For this measurement nails are driven through the board at right angles to the face with the nailhead set flush with the surface of the board. The specimen is placed in a holding fixture with the pointed end of the nail in a tension grip or "Jacobs-type drill chuck" which is attached to the head of a testing machine. The maximum load required to pull the head of the nail through the board is a measure of its resistance to nailhead pull-through.

Direct Screw Withdrawal[1]

The maximum load required to withdraw a screw from a board either perpendicular to the plane of the board or from the edge of the board is determined in this test. The manner in which the screw is to be inserted is carefully controlled.

DIMENSIONAL STABILITY PROPERTIES

Water Absorption and Thickness Swell[1]

Square specimens, after standard conditioning, are accurately measured by a standardized procedure for width, length, and thickness, and the volume calculated. After being accurately weighed, the specimens are submersed in room-temperature water for 2 hours. After removal, the specimens are allowed to drain a specified time, excess water is blotted from the surfaces, and the specimens are weighed and their thickness measured. They are again submersed in room-temperature water for an additional 22-hour period, and the draining, weighing, and measuring procedures repeated. The specimens are then dried in an oven at a temperature slightly above the boiling point of water until they reach constant weight. From these data the amount of water absorbed after conditioning and absorbed during the two submersion periods can be calculated as a percentage by volume or by weight based on volume and the ovendry weight. Thickness swelling is expressed as a percentage of the original thickness.

Edge Thickness Swelling by the Disk Method[1]

The use of a small disk specimen permits the measurement of the maximum edge swelling that might be obtained. In the test for thickness swelling described above, the method does not yield the maximum value possible. The disk method employs specimens of 1 inch diameter by the thickness of the board, which allows more rapid and uniform water absorption when submersed than can take place in the panel test previously described.

Linear Variation With Change in Moisture Content[1]

The dimensional stability of a board as affected by moisture content changes is determined by measurement of variations of length. Specimens are cut from boards with the test length of the specimen cut parallel with the long dimension of the board and others from the same board at right angles to the long dimension. The specimens are cut narrow in width to accelerate attainment of an equilibrium moisture content. After conditioning at 50 percent relative humidity, the specimens are accurately measured for length. Although any of a variety of moisture condition changes can be evaluated, the standard recommends placing some specimens in a 90 percent relative humidity environment and others in a 30 percent relative humidity until equilibrium moisture content is reached. An accurate measurement of length made after the specimens are at equilibrium moisture content, allows calculation of the linear change that occurs between 50 and 90, and 50 and 30 percent relative humidity. The change in length between 30 and 90 percent relative humidity may be estimated with sufficient accuracy by simple addition of the first two values.

Cupping and Twisting[1]

These distortions can be measured on either specimens or boards that have been subjected to exposures involving moisture content variations. Cupping is determined by placing a straight-edge across opposite edges of a specimen and measuring the maximum distance to the concave face. Twisting may be determined by placing the specimen with three corners touching a level surface and measuring the distance from the raised corner to the level surface.

PERFORMANCE PROPERTIES

The measurements described above, in general, used small panels or small-sized specimens for tests. Those to be described here involve, in most cases, larger panels, larger areas, or component assemblies. These are often proof tests of composites that were designed on the basis of values obtained from small specimen testing as, for example, racking tests in shear walls. Others such as fire

performance tests require large areas for proper comparison with
other materials. The only small-specimen tests to be included as a
performance test are those for evaluating formaldehyde emissivity
from products bonded with formaldehyde-containing resins.

Fire Performance

The flame-spread properties of wood composites are measured by a
so-called 25-foot tunnel furnace test.[19] This test determines compara-
tive burning characteristics by determining the rate of flame-spread
over the surface of materials when exposed to a test fire under controlled laboratory conditions. Spread rate values are used to
develop surface flammability ratings. Smoke density values are also
obtained. The ability of an assembly to act as a barrier to fire is
determined by a flaming exposure that is controlled to a time-
temperature regimen.[20] Load-carrying ability during exposure can be
determined as a factor contributing to fire containment or to
maintenance of structural integrity.

Accoustical Performance

A single-figure rating or index can be determined[21] regarding
the insulation provided against the sounds of speech, radio, tele-
vision, music, and similar sources of noise in offices and buildings.
The method can be used to compare partitions designed for general
building purposes. In another method, sound-insulating properties
of a partition are expressed in terms of the sound-transmission loss,
which refers to specimens exposed to a diffuse incident sound field.[22]
A diffuse sound field test procedure for evaluating the sound-
insulating properties of building elements, as well as other test
procedures, has also been standardized. A correlation between
laboratory-derived data and field tests on sound insulation properties of materials and construction variables has been developed by
Jones.[24]

Structural Performance

A variety of procedures[9] is available to determine the structural properties of segments of wall, floor, and roof constructions.
These involve deformation and strength measurements under different
loading situations including compression, tension, transverse,
concentrated, impact, and racking loads.

Formaldehyde Emissivity

A performance property of wood composites that has become
increasingly important in recent years involves the tendency for
formaldehyde release from composites bonded with urea-formaldehyde
resin adhesives. The formaldehyde concentrations in living spaces,

where such materials have been installed, may become uncomfortable
or possibly unhealthy under some adverse conditions.

A wide variety of procedures have been employed to characterize
small specimens for their formaldehyde-emitting potential and there
is, as yet, no universally accepted method. The so-called free or
chemically unbound formaldehyde can be estimated using the Perforator
extraction procedure,[25] and this has been widely employed in some
European countries. In the United States two procedures which are
receiving serious attention as possible standards are modifications
of the Equilibrium Jar test, originally developed by Georgia-Pacific
Corporation,[26] and the Japanese Desiccator test.[27,28] Measurements
of formaldehyde concentration in an air-purged system can provide
information which is more directly related to actual use conditions.[28]

In all of the above the formaldehyde is obtained either directly
as a water solution or indirectly after scrubbing a given volume of
air into water. The aqueous solution can then be analyzed for its
formaldehyde content using colorimetric of fluorometric methods with,
for example, chromotropic acid[26,29] or acetylacetone[28] as reagents.

DURABILITY PROPERTIES

Since wood composites are intended for long-term use in a wide
variety of structural applications in diverse service environments,
durability properties are of prime importance. But the evaluation
of durability properties is a very challenging and complex problem.
Procedures for evaluation are continuously evolving and new ones
being developed. Confidence levels regarding wood composite durability increase with the progressive accumulation of test results
from laboratory exposures and weathering studies which are designed
to evaluate different aspects of the problem. The long-term nature
of intended uses dictates the need for accelerated aging procedures
that may be extrapolated meaningfully to the long term. Here, there
is a need to compare the performance of new composites with that of
materials known to provide satisfactory service.

There are essentially three stages in durability assessment,
each with somewhat different objectives, that cumulatively provide
the confidence that new wood composites can be expected to perform
satisfactorily with a reasonably long service life. The first stage
concentrates on an assessment of the durability properties of adhesives so they may be properly selected and used for some intended
purpose. The second stage concentrates on an evaluation of wood
composite properties. A series of tests is required to determine
whether or not a new wood composite will have the expected performance when properly manufactured with the selected adhesive. In the
third stage, short-term tests are required for quality assurance that
the wood composite is properly manufactured during plant operations.

A number of factors contribute to the degradation processes that adhesives and wood composites may undergo. These degrading influences can be either the chemical effects of aging or the physical effects of imposed loads, or often combinations of both effects.

Among the factors that contribute to the chemical effects of aging are the effects of heat, chemicals, and microorganism attack, which is a special case of chemical effects. The chemical effects involve an array of materials which can degrade the adhesive, the wood substrate, or the interface. The most common chemical influence is that caused by water which is present in all wood-based products in almost every service environment. Other chemical effects might be caused by the natural constituents of wood, the ingredients of the adhesive, preservative or fire-retardant treatments, or chemicals present in the service environment.

The physical effects of imposed loads may be due to internal stresses from swelling and shrinking or may be due to an externally applied load either cyclical or of short-term duration, or continuous long-term duration.

Before they can be used with confidence, new adhesive systems must be evaluated for the above effects, either as single effects, as much as possible, or as combinations.

Evaluating Adhesives as Candidates for Long-Term Performance

Determining the long-term performance of new adhesive systems requires accelerated aging in a manner that allows some rational means of extrapolating to long-term service conditions. Thermal degradations and those caused by chemicals are temperature-dependent processes and can be evaluated at elevated temperatures, and the rates of strength loss at service temperatures can be estimated by the Arrhenius temperature-dependent relationship.[30] Determining the rates of strength loss of bonded specimens at different elevated temperatures under essentially dry heat conditions evaluates thermal effects alone. Most adhesives show life expectancies of hundreds or thousands of years at room temperatures when dry, just as wood is known to behave historically.[31]

The effects of moisture can be determined in a similar manner, maximizing the effect by determining rates of strength loss under water-soak conditions.[30] Strength loss rates determined under dry and water-soaked conditions represent the two extremes with rates obtained at intermediate moisture contents ranging between these extremes.[32] The results of 8 to 10 years weathering of materials previously evaluated by this method of accelerated aging showed that a satisfactory correlative ranking existed between accelerated aging and weathering, and that accelerated aging was able to detect causes of degradation not elucidated by weathering studies.[33] The precision of the rate-process method of accelerated aging depends upon the number of specimens tested, the number of test temperatures, and the extent of extrapolation. While a broad error band is associated with such projections, the use of the lower 95 percent confidence limits provides a positive means for making reasonable estimates of minimum bond line durabilities.[34] The interaction of other chemicals on bond line durability can be determined by similar rate-process techniques as has been demonstrated for the effects of acid catalysts used for curing adhesives.[31] In such cases, water soaking may reduce the rate of degradation by leaching the responsible chemical away from the bond line.

The chemical degradation caused by microorganisms cannot be accelerated by use of elevated temperatures because they grow and fluorish only within a narrow temperature range. These effects have been evaluated by ASTM standard procedures.[35]

The physical effects caused by loads imposed on bonded joints are reflected in the changes that occur in the important mechanical properties in service environments. These are the same mechanical properties used in engineering design of structural components. In most cases structures are designed so that bonded joints become loaded in the shear mode, although at times a tensile mode must be taken into consideration. Consequently, the important properties requiring measurement include shear strength and shear modulus, which provide the strength and stiffness requirements for short-term loads, and creep resistance and duration of load or endurance limit properties, which are required for continuous long-term loads.

Accurate methods are available for measuring shear strength and shear modulus of adhesives in bonded joints including those adhesives that are less strong and stiff than wood itself. These methods include a thick-adherend lap-shear test,[36,37] and a modified rail test.[38] Creep, creep relaxation, and endurance limit values can be obtained by the classical Wöhler method using continuous dead loads.[39] Duration of load or an endurance limit, the load that a joint can sustain continuously without failure, may also be estimated by the Prot method which evaluates specimens under progressively reduced rates of loading, followed by extrapolation of the failure loads to zero rate of loading.[40]

A method has been developed to evaluate how different adhesives respond to internal stress development as wood substrates change dimensions with moisture changes. The adhesives are evaluated in crossed-grain hard maple specimens for testing in compression shear after equilibrating to different moisture contents.[41]

Thermal analysis such as differential scanning calorimetry (DSC), differential thermal analysis (DTA), and thermogravimetric analysis (TGA), along with spectroscopic analysis can be used to determine chemical structural relationships in thermosets that affect subsequent durability. Chow et al. [42,43] developed a measurement of a thermal softening temperature for thermosets probably related to the glass transition temperature. They demonstrated that adhesive resins with higher thermal softening temperature were, in general, of higher durability. Chow et al. [44-48] also used thermal analysis and infrared spectroscopy to relate resin-preparation and resin-curing variables to the durability performance of phenolic-type adhesives. Nachtrab,[49] Schindlbauer et al.,[50] and Rosenberg[51] also used thermal analysis techniques to evaluate how temperature and time differences affected the degree of cure in phenolics.

Evaluation of New Wood Composites for Long-Term Performance

When a new bonded wood product is developed for long-term performance requiring a durable adhesive, the choice of adhesive is made on the basis of its past history of successful use, or from the information generated by the tests described above. Consequently, the adhesive is known to have high resistance to all factors influencing degradation. However, it must still be demonstrated that the new product properly bonded with specified amounts of the adhesive will actually perform as intended. Differences in performance depend largely upon the amount of adhesive used and how well the bonds resist internal stress development. Consequently, accelerated-aging tests are designed to maximize internal stress development, and the rate of loss of some important end-use property is determined during repeated cycles of swelling and shrinking.

A case in point is the evaluation of structural flakeboards from forest residues for which phenolic adhesives were selected as the binder.[52] Here the bending strength, bending stiffness, internal bond strength, and thickness properties were determined throughout multiple cycles of boiling in water and elevated-temperature drying. Elevated temperatures could be used since the adhesive was known to be resistant to thermal and moisture effects, and it was desired that the rate of change in swelling and shrinking be maximized and time to carry out each cycle be minimized.

The results of such accelerated aging can be interpreted as how much of each important property is retained even after severe treatment upon comparison with how conventional materials of known satisfactory performance behave when subjected to the same cyclic treatment. It was concluded that phenolic-bonded flakeboards containing at least 5 percent resin would be expected to perform satisfactorily in structural applications where conventional wood products are now used.

The use of elevated temperatures in multiple-cycle aging is not universally accepted. There is some preference for a cycle consisting of a vacuum-pressure-soak at tap water temperature with drying at no higher than 82° C (180° F) even though cycle time is greatly increased. Multiple-cycle aging under these conditions has been used to evaluate red oak structural particleboard in comparison with other materials commonly used for industrial or commercial roof decking.[53]

The above approach to the durability evaluation of new wood composites is a departure from that customarily used. The usual approach is to devise laboratory procedures for exposing specimens to some selected conditions normally involving cyclic swelling and shrinking and determine how some important mechanical property changes during these exposures. Attempts are then made to correlate these results with those obtained in weathering studies. Efforts are made to deduce the weathering conditions influencing degradation and to simulate these conditions in the laboratory as much as possible and practicable.

There are numerous problems associated with this approach: in selecting the weathering factors responsible for degradation, in simulating an appropriate mix and intensity of these factors, in selecting climates typical of service environments, in coping with seasonal and annual fluctuations in climates, and in acclerating laboratory treatments to reduce testing time. In spite of the many problems, important and significant progress in developing useful wood composites has been made by this approach.

One of the first accelerated-aging procedures to be developed for evaluating wood-based materials was standardized as an ASTM method[1] and is still widely used today. The exposures provide a complex mixture of soaking, steaming, freezing, and drying through 6 cycles requiring 12 days. Although this procedure required a lengthy testing period, it became a quality assurance test in the United States for mat-formed wood particleboard (CS 236-66).[54] Replacement of this procedure with a less time-consuming but still discriminatory test is badly needed. The West Coast Adhesive Manu-facturer's Association (WCAMA)[55,56] proposed to reduce the time required for this test with a 6-cycle procedure of vacuum-pressure soaking, boiling, and drying, but reduced the time by only half. More recently, Lehmann[57] evaluated a number of different cyclic moisture conditions on the strength and stability of flakeboards. From the results obtained, Lehmann advocated the use of only moderate temperatures during water exposure or drying limited to the maximum expected in service. Materials, when subjected to high temperatures, showed high rates of strength loss which might make them appear less durable than would be the case in actual service.

Many wood composites contain bond lines below the depths pene-
trated by ultraviolet radiation. Others, such as fiberboards,
particleboards, and flakeboards have bond lines extending to the
surface or protected by only thin layers of wood. In such cases,
ultraviolet radiation may play a role in the degradation process and
some investigators have used ultraviolet light in their cyclic procedures. Deppe and Stolzenburg[58] used a specially designed BAM
(Bundesanstalt für Materialprüfung) weather tunnel to evaluate the
exterior durability of particleboards including coated materials.
The technique involved a complex series of cycles of water spray or
immersion, heat storage, UV irradiation, and cold storage. In a
subsequent study,[59] they determined that the BAM weather tunnel gave
results in 3 months comparable with 5 years of weathering.

Deppe and Stolzenburg[60] later extended these evaluations to
include another weathering chamber and compared the results with
those from other short-term tests as well as natural weathering.
Their results to date suggest that UV irradiation is an important
factor and weathering chambers provide results that parallel natural
weathering more closely than do the more conventional cyclic
procedures.

In contrast, Beech et al. [61,62] compared several of the common
short-term tests with a specially made "weatherdrum" which incorporated water spray, UV irradiation, and drying cycles. They found
that a cyclic soak-freeze-thaw procedure (test designation V-313)
gave the best correlation with 2-year natural weathering as measured
by bending strength response. The "weatherdrum" provided no better
predictive capability in this case.

A continuing effort is needed to determine the most practical
mix of degrading influences and their intensities that will correlate well with natural weathering. Weathering chambers should prove
useful, if their cycles can be "tuned" adequately for good correlation,
for they have the capability for complete automation of the cycles.
Many of the proposed cyclic procedures do not lend themselves to
automation but require specimens to be moved from one apparatus to
another manually. This severely limits most treatments to progress
only during working hours rather than 24 hours a day, including
weekends and holidays. The desire to automate cyclic treatments,
but omitting any UV irradiation, led to the development of an automatic machine for boil-dry cycles at the Canadian Forest Products
Laboratory in Vancouver.[63] This machine was further refined and
extensively evaluated by the Weyerhaeuser Company[64] and later adopted
as a standard recommended practice.[65]

In service, structural elements of buildings are subjected to
moisture and temperature variations while under continuous or inter-
mittent loads. However, there is only limited information available

about the changes that take place in the mechanical properties of wood composites while under load in the variable climates found in service. One of the most extensive studies was reported by Gressel[66] who investigated the bending properties of particleboard, plywood, and solid wood as affected by moisture content, temperature, and bending stress variations. It was found that particleboards exhibited higher absolute and relative creep rates in a climate of high humidity than in a climate of varying humidity: a performance opposite to the manner in which plywood and solid wood behaved. There is a need for additional studies of durability properties that combine the chemical effects of aging with the physical effects of loading.

Quality Assurance Evaluation of Wood Composites

Test procedures to assure control over the manufacturing process should require only a few hours or a day or two to carry out and should readily discriminate between properly manufactured products and those that are not. Such tests have usually evolved through extensive efforts to correlate the results of some arbitrarily selected short-term laboratory treatments with the results of weathering exposures. However, if a thorough study has been made of a new product's expected long-term performance as described in the previous section, the background information obtained can provide an excellent starting point for developing short-term quality assurance tests. For example, in the multiple-cycle soak-dry evaluations described earlier[52,53] to compare the behavior of a new product with that of materials with known durability, it can be concluded that as few as five cycles should be sufficient for a quality assurance test.

Quality assurance tests also evolve in laboratories developing new raw materials or new products in response to their need for rapid screening tests. Although such laboratories usually use a standard quality control test for screening purposes, such tests are often time-consuming and cumbersome and pressures arise for modification and simplification.

Many different standard procedures are used by the wood-composite industries in the United States. For example, the construction-grade plywood standard[67] describes a boil-dry-boil cycle and a vacuum-pressure-soak cycle to evaluate exterior-type bonds. A similar boil-dry-boil cycle for exterior-type bonds can be found in the hardwood and decorative plywood standard[68] along with a two and a three soak-dry cycle procedure for interior-type bonds. The standard for structural glue laminated[69] timber describes a vacuum-pressure soak-dry procedure for delamination measurement following a method in the inspection manual[70] of the American Institute of Timber Construction. However, as mentioned earlier, the quality-control test for particleboard and flakeboard[54] follows a lengthy and complicated procedure.[1]

A German standard[71] for construction-grade particleboard describes
a 2-hour boil followed by wet testing for internal bond strength.
This exposure is also described in the Canadian standard[72] but applied
to bending specimens. Shen and Wrangham[73] reported on a modification
of the 2-hour boil test wherein the internal bond strength was correlated with a torsion shear test. Most recently Clad[74] has proposed
a 5-minute vacuum-soak test with shear strength measurements as a means
to determine quality differences among various particleboards.

There is a continuing need for improvements in quality control
testing, to reduce cost and time requirements, and to improve their
capability to detect important differences in product quality.

POSTSCRIPT

The author has made an attempt to provide an overview of the
current status of a most complex subject: to describe the critical
properties of wood composites and show how they are measured. Little
more than a superficial treatment of each aspect of the problem has
been presented. Some important and significant work has been omitted,
some inadvertently, some due to the author's undoubtedly incomplete
knowledge, and some due to individual biases. A somewhat cursory
attempt has been made to present different points of view, different
philosophies of approach, and different interpretations of results.
If this summary of the current state-of-the-art serves only to
stimulate further thought, discussions, and new research studies, it
will have served its purpose.

LITERATURE CITED

1. American Society for Testing and Materials. Standard methods of
     evaluating the properties of wood-base fiber and particle
     panel materials. ASTM Desig. D-1037-78, ASTM, Philadelphia,
     Pa. (1979).
2. American Society for Testing and Materials. Standard methods of
     testing plywood in tension. ASTM Desig. D-3500-76, ASTM,
     Philadelphia, Pa. (1979).
3. American Society for Testing and Materials. Standard methods of
     testing plywood in compression. ASTM Desig. D-3501-76, ASTM,
     Philadelphia, Pa. (1979).
4. American Society for Testing and Materials. Standard test method
     for strength properties of adhesive bonds in shear by compres-
     sion loading. ASTM Desig. D-905-49 (Reapproved 1976). ASTM,
     Philadelphia, Pa. (1979).
5. American Society for Testing and Materials. Standard method of
     testing plywood in rolling shear (shear in plane of plies).
     ASTM Desig. D-2718-76, ASTM, Philadelphia, Pa. (1979).

6. American Society for Testing and Materials. Standard method of
    test for shear modulus of plywood. ASTM Desig. D-3044-76,
    ASTM, Philadelphia, Pa. (1979).
7. American Society for Testing and Materials. Standard method of
    testing plywood in shear through-the-thickness. ASTM Desig.
    D-2719-76, ASTM, Philadelphia, Pa. (1979).
8. American Society for Testing and Materials. Standard test
    method for toughness of plywood. ASTM Desig. D-3499-76,
    ASTM, Philadelphia, Pa. (1979).
9. American Society for Testing and Materials. Standard methods of
    conducting strength tests of panels for building construction.
    ASTM Desig. E-72-77, ASTM, Philadelphia, Pa. (1979).
10. American Society for Testing and Materials. Standard methods of
    testing plywood in flexure. ASTM Desig. D-3043-76, ASTM,
    Philadelphia, Pa. (1979).
11. J. E. Gordon. The new science of strong materials. Penguin
    Books, New York, N.Y. (1976).
12. Y. Lei and J. B. Wilson. Fracture toughness of oriented flake-
    board. Wood Sci. 12(3):154-161 (1980).
13. M. S. White and D. W. Green. Effect of substrate on the frac-
    ture toughness of wood-adhesive bonds. Wood Sci. 12(3):
    149-153 (1980).
14. R. Ebewele, B. River, and J. Koutsky. Tapered double cantilever
    beam fracture tests of phenolic-wood adhesive joints. Part I.
    Wood and Fiber 11(3):197-213 (1979).
15. J. D. McNatt. Effect of rate of loading and duration of load
    properties of particleboard. For. Serv. Res. Pap. FPL-270.
    Forest Products Laboratory, Madison, Wis. (1975).
16. M. Kufner. Creep in wood particleboard under long-term bending
    load. Holz als Roh- und Werkstoff 28(11):429-446 (1970).
17. R. G. Pearson. The effect of duration of load on the bending
    strength of wood. Holzforschung 26(4):153-157 (1970).
18. C. C. Gerhards. Effect of duration and rate of loading on
    strength of wood and wood-based materials. For. Serv. Res.
    Pap. FPL-283. Forest Products Laboratory, Madison, Wis.
    (1977).
19. American Society for Testing and Materials. Standard method of
    test for surface flammability of building materials. ASTM
    Desig. E-84-79, ASTM, Philadelphia, Pa. (1979).
20. American Society for Testing and Materials. Standard methods
    of fire tests of building construction and materials. ASTM
    Desig. E-119-79, ASTM, Philadelphia, Pa. (1979).
21. American Society for Testing and Materials. Standard classifi-
    cation for determination of sound transmission class. ASTM
    Desig. E-413-73, ASTM, Philadelphia, Pa. (1979).
22. American Society for Testing and Materials. Standard method for
    laboratory measurement of airborne sound transmission loss of
    building partitions. ASTM Desig. E-90-75, ASTM, Philadelphia,
    Pa. (1979).

23. American Society for Testing and Materials. Standard method of
    test for measurement of airborne sound insulation in buildings.
    ASTM Desig. E-336-77, ASTM, Philadelphia, Pa. (1979).
24. R. E. Jones. Effects of flanking and test environment on lab-
    field correlations of airborne sound insulation. J. of the
    Accoustical Society of America, JASA 57(5):1138-1149 (1975).
25. Federation Européenne des Syndicats de Fabricants de Panneaux de
    Particules (FESYP). Determination of formaldehyde emitted
    from particleboard (1975).
26. Georgia Pacific. Georgia Pacific Analytical Method 203.6.
    Colorimetric determination of formaldehyde. Georgia Pacific
    Corp., Decatur, Ga. (1979).
27. Japanese Industrial Standard A5908.
28. G. E. Myers and M. Nagaoka. Formaldehyde emission: methods of
    measurement and effects of particleboard variables. Submitted
    to For. Prod. J. (1980).
29. National Institute of Safety and Health (NIOSH). A recommended
    standard for occupational exposure to formaldehyde, Appen-
    dix I (1977).
30. R. H. Gillespie. Accelerated aging of adhesives in plywood-type
    joints. For. Prod. J. 15(9):369-378 (1965).
31. R. H. Gillespie and B. H. River. Durability of adhesives in
    plywood: dry heat effects by rate-process analysis.
    For. Prod. J. 25(7):26-32 (1975).
32. R. H. Gillespie. Parameters for determining: heat and moisture
    resistance of a urea-resin in plywood joints. For. Prod. J.
    18(8):35-41 (1968).
33. R. H. Gillespie and B. H. River. Durability of adhesives in
    plywood. For. Prod. J. 26(10):21-25 (1976).
34. M. A. Millett and R. H. Gillespie. Precision of the rate-process
    method for predicting bondline durability. Forest Products
    Laboratory. Prepared for Department of Housing and Urban
    Development (HUD). National Technical Information Service
    (NTIS) PB80-121866 (1978).
35. American Society for Testing and Materials. Effect of mold
    contamination on permanence of adhesive preparations and
    adhesive bonds. ASTM Desig. D-1286. Permanence of adhesive-
    bonded joints in plywood under mold conditions. ASTM Desig.
    D-1877. Effect of bacterial contamination on permanence of
    adhesive preparations and adhesive bonds. ASTM Desig. D-1174.
    ASTM, Philadelphia, Pa. (1979).
36. B. H. River and R. H. Gillespie. Measurement of shear modulus
    and shear strength of adhesives. Forest Products Laboratory.
    Prepared for Department of Housing and Urban Development (HUD).
    National Technical Information Service (NTIS) PB80-121742
    (1978).

37. B. H. River. Strength and shear moduli of several construction adhesives as influenced by environment and loading conditions. Forest Products Laboratory. Prepared for Department of Housing and Urban Development (HUD). To be published by National Technical Information Service (NTIS) (1978).

38. G. P. Kreuger. Determination of shear properties of flexible structural adhesives by the modified rail test. Submitted to American Society for Testing and Materials (ASTM) for consideration as a standard recommended practice. Institute of Wood Research, Michigan Technological University, Houghton, Mich. 49931 (1975).

39. B. H. River and R. H. Gillespie. Long-term load-deformation properties of adhesives. Forest Products Laboratory. Prepared for Department of Housing and Urban Development (HUD). To be published by National Technical Information Service (NTIS) (1978).

40. F. H. M. Nestler. The strength of adhesive-bonded wood products: evaluation of the Prot method of progressive loading. (Under review for publication as USDA FS-FPL research paper.) (1979).

41. R. H. Gillespie. Effect of internal stresses on bond strength of wood joints. Prepared for Department of Housing and Urban Development (HUD). NTIS-PB-258832/5WB.

42. S. Chow. Softening temperatures and durability of wood adhesives. Holzforschung 27:64-68 (1973).

43. S. Chow and R. W. Caster. Relationship of adhesive softening temperature to exposure tests for bond durability. For. Prod. J. 28(6):38-43 (1978).

44. S. Chow. Thermal analysis of liquid phenol-formaldehyde resin curing. Holzforschung 26:229-232 (1972).

45. S. Chow. A curing study of phenol-resorcinol-formaldehyde resins using infrared spectrometer and thermal analysis. Holzforschung 31:200-204 (1977).

46. S. Chow and P. R. Steiner. Comparisons of the cure of phenol-formaldehyde novolac and resol systems by differential scanning calorimetry. J. Appl. Polym. Sci. 23:1973-1985 (1979).

47. S. Chow and P. R. Steiner. Comparison of curing and bonding properties of particleboard and waferboard-type phenolic resins. For. Prod. J. 29(11):49-55 (1979).

48. S. Chow, P. R. Steiner, and G. E. Troughton. Thermal reactions of phenol-formaldehyde resins in relation to molar ratio and bond quality. Wood Sci. 8(1):343-349 (1975).

49. G. Nachtrab. Investigation of the degree of cure of thermosetting resins using differential thermal analysis. Kunststoffe 60:261-265 (1970).

50. H. Schindlbauer, G. Henkel, J. Weiss, and W. Eichberger. The qualitative study of curing behavior of phenoplasts during differential thermal analysis measurements. Angew. Makromol. Chem. 49(725):115-128 (1976).

51. G. N. Rosenberg. Thermal softening as a method of determining resin cure. Holzforschung 32(3):92-95 (1975).

52. A. J. Baker and R. H. Gillespie. Accelerated aging of phenolic-bonded flakeboards. USDA-FS Gen. Tech. Rept. WO-5, pp. 93-100 (1978).

53. M. O. Hunt, W. L. Hoover, D. A. Fergus, W. F. Lehmann, and J. D. McNatt. Red oak structural particleboard for industrial/commercial roof decking. Report RB 954, Wood Research Laboratory, Dept. of Forestry and Natural Resources, Agric. Exp. Sta., Purdue University, West Lafayette, Ind. 47907 (1978).

54. U.S. Department of Commerce, National Bureau of Standards. Mat-formed wood particleboard. Commercial Standard CS-236-66. U.S. Department of Commerce, Washington, D.C. (1966).

55. West Coast Adhesive Manufacturers Association Technical Committee. A proposed new test for accelerated aging of phenolic-resin bonded particleboard. For. Prod. J. 16(6): 19-23 (1966).

56. West Coast Adhesive Manufacturers Association Technical Committee. Accelerated aging of phenolic-resin bonded particleboard. For. Prod. J. 20(10):26-27 (1970).

57. W. F. Lehmann. Cyclic moisture conditions and their effect on strength and stability of structural flakeboards. For. Prod. J. 28(6):23-31 (1978).

58. H.-J. Deppe and R. Stolzenburg. Coated particleboards for exterior building work. Holz-Zbl. 97(19):247-248 and 97(25):349-350 (1971).

59. H.-J. Deppe and R. Stolzenburg. Accelerated weathering tests to assess the durability of coated composite wood panels. Holz-Zbl. 101(83):1084-1087 and 101(128):1675-1678 (1975).

60. H.-J. Deppe, R. Stolzenburg, and K. Schmidt. Testing and evaluating the durability of wood particleboard by means of short-term weathering processes. Holz Roh-Werkstoff 34: 379-384 (1976).

61. J. C. Beech, R. W. Hudson, R. A. Laidlaw, and L. C. Pinion. Studies on the performance of particleboard in exterior situations and the development of laboratory predictive tests. Building Research Establishment Current Paper CP 77/74 (1974).

62. R. A. Laidlaw and J. C. Beech. The assessment of predictive tests for forecasting the performance of particleboard in exterior environments. Proc. of IUFRO-5 Meeting, Vol. 2, Republic of South Africa, pp. 620-633 (1973).

63. D. C. Walser and H. G. M. Colbeck. Bond-degrade accelerating machine helps predict bond life. Adhesives Age 10(11): 33-35 (1967).

64. R. E. Kreibich and H. G. Freeman. Development and design of an accelerated boil machine. For. Prod. J. 18(12):24-26 (1968).

65. American Society for Testing and Materials. Standard recommended
     practice for multiple-cycle accelerated-aging test (automatic-
     boil test) for exterior (wet-use) wood adhesive. ASTM Desig.
     D-3434-75, ASTM, Philadelphia, Pa. (1979).

66. P. Gressel. The effect of time, climate, and loading on the
     bending behavior of wood-base materials. Holz Roh-Werkstoff
     30:259-266; Part II. Test results dependence on the creep
     parameters. 30:347-355; Part III. Discussion of results.
     30:479-488. (1972).

67. U.S. Department of Commerce, National Bureau of Standards. Con-
     struction and industrial plywood. Voluntary Products Stan-
     dard PS-1-74, U.S. Department of Commerce, Washington, D.C.
     (1974).

68. U.S. Department of Commerce, National Bureau of Standards.
     Hardwood and decorative plywood. Voluntary Products Stan-
     dard PS-51-71, U.S. Department of Commerce, Washington, D.C.
     (1971).

69. U.S. Department of Commerce, National Bureau of Standards.
     Structural glued laminated timber. Voluntary Products Stan-
     dard PS-56-73, U.S. Department of Commerce, Washington, D.C.
     (1973).

70. American Institute of Timber Construction. Inspection Manual
     AITC-200-73, American Institute of Timber Construction,
     333 West Hampden Avenue, Englewood, Colo. 80110 (1973).

71. German Standard. Particleboards; flat-pressed boards for
     building; terms, properties, testing, supervision. DIN 68763,
     Deutsche Normenausschusses, Berlin (1973).

72. Canadian Standards Association. Mat-formed wood particleboard.
     CSA-0188-1975, Canadian Standards Association, Rexdale,
     Ontario (1975).

73. K. C. Shen and B. Wrangham. A rapid accelerated-aging test
     procedure for phenolic particleboard. For. Prod. J. 21(5):
     30-33 (1971).

74. W. Clad. Testing of particleboard after short evacuation in
     water. Holz Roh-Werkstoff 37:419-425 (1979).

TIME DEPENDENT PROPERTIES

Ronald G. Pearson

Department of Wood and Paper Science
North Carolina State University
Raleigh, North Carolina 27650

INTRODUCTION

As described in earlier papers, wood is an oriented composite
of remarkable complexity. Connections between the structural components range from the numerous strong hydrogen bonds in the
crystalline regions of the microfibrils through a limited number
of covalent bonds between the many-branched lignin and adjacent
carbohydrate polymers (probably hemicellulose) to the relatively
sparse hydrogen bonds in the amorphous regions.

Although not regarded as part of the woody substance, the
fiber walls contain water which plays a vital role in determining
the performance of wood in service and its behavior under stress.
Green wood has saturated cell walls, but below fiber saturation
point the amount of water varies with the relative humidity of the
ambient air.

The addition of an adhesive to wood to form a man-made composite product introduces many new possible sources of variation in
properties and response to stress. The species and type of wood
component (veneer, flakes, fibers, particles, laminations), the
type, formulation and quantity of resin, the degree of coverage of
the wood surface by the resin, the orientation of the wood grain,
the temperature, pressure and other process variables are among the
parameters affecting the properties. Consequently, it is not surprising that the study of time-dependent properties is still mainly at the phenomenological level. Although there are considerable
differences between the behavior of solid wood and wood composites,
the time-dependent properties of solid wood will be discussed in
some detail to provide a foundation for the understanding of the
performance of composites. Also included in this discussion,

although not strictly time dependent, are the effects of moisture
change which are so intimately linked to the behaviour of
stressed wood.

EFFECT OF DURATION OF STRESS ON STRENGTH

Solid Wood

   The standard procedure for determining the strength of wood is
to increase the load continuously (ramp loading) under specified
conditions until failure occurs. The rate of loading is such as
to produce failure within 5 to 15 minutes. Faster rates of loading
lead to higher failing loads, slower rates lead to lower loads at
failure. It is impractical to apply load continuously at very slow
rates of loading, so the effect of long duration of stress is
usually determined by loading to some percentage öf the load estimated to cause failure in the standard test and then maintaining
that level of stress until failure occurs--a "constant load" test.

   Wood (1951) fitted the hyperbolic expression given in equation
(1) to the results of rate of loading and constant load tests on
Douglas fir. This expression, which is commonly called the
"Madison formula" and is illustrated in Figure 1, is used in North
America for allowing for the effects of duration of load in the
derivation of allowable stresses for structural design.

$$SL = 18.3 + 74.34 \; t^{-0.04635} \tag{1}$$

where SL = failing stress as percentage of short time ultimate
           stress as determined from a standard test
      t = time from start of test (hr).

   The hyperbolic expression was chosen partly because the rate
of loading tests indicated a larger increase in failing stress than
that obtained by extrapolation of the constant load test results
and partly because it seemed reasonable to assume the existence of
a threshold stress below which failure would never occur.

   Experimental results have also been well fitted by a semi-
logarithmic expression. Pearson (1972) obtained equation 2 as a
fit to all rate of loading and constant load test results published
up to that time, 20% of the total time of the rate of loading tests
being arbitrarily taken as the effective duration of those tests.

$$SL = 91.5 - 7 \; \log_{10}t \tag{2}$$

where t = time under load (hr).

   Equation (2) does not provide for a threshold of stress, and none was indicated by the test results, which had durations ranging up to nearly 10 years. Despite differences in species, size and moisture content, there was a marked agreement on the general effect of load on strength. For a given stress level the duration to failure covered three orders of magnitude, and, as may be seen from equation (2), a 7% reduction in the stress level increased the average time to failure ten-fold.

Failure of clear wooden beams commences in the compression zone with the development in the fiber walls of microscopic slip planes which enlarge into microscopic "wrinkles" as the stress increases (Dinwoodie 1968). Keith (1971) reported the presence of such microscopic compression failures in white spruce specimens loaded in compression parallel to the grain. They were noticeable in quantity only for stresses exceeding 60% of the estimated short-term failing stress, and developed earlier in specimens maintained at 18% moisture content than in those maintained at 9% moisture content. Armstrong and Kingston (1968) observed minute compression failures in compression specimens and in beams drying under load. None were observed in beams maintained at constant moisture content, probably because the stress levels did not exceed 40% of the short-term strength.

Schniewind and Centeno (1973) examined the effect of duration of load on the fracture toughness of Douglas fir with cracks oriented in the six principal planes of symmetry. The bending strength of wood stressed perpendicular to the grain was studied at various stress levels by Bach (1975) and at various deformation rates by Mindess et al. (1976). Bach obtained an equation similar to that of equation (2) but with a steeper gradient while Mindess and his colleagues concluded that time-dependent failure was governed by crack kinetics.

Several criteria for strength under duration of load have been proposed. Ylinen (1957) proposed a maximum strain approach. Bach (1973) obtained promising results with the Reiner-Weiseinberg criterion that the total energy stored visco-elastically at failure is constant. Schniewind (1979) presented a review of the effects of duration of load on solid wood and discussed possible mechanisms of failure.

Gerhards (1977, 1979) introduced damage theory similar to that used to describe accumulating fatigue or creep rupture strength in metals to provide a model for duration of load effects on the strengths of wood. This work was extended by Barrett and Foschi (1978). They assumed that the rate of damage was proportional to the stress level and to the amount of damage already suffered by the wood due to its past loading history. The model was fitted to some constant load data to obtain values of the parameters. With these parameters, the model was then found to provide a good fit

to ramp loading data as illustrated in Figure 1. Barrett and
Foschi concluded that "the mechanisms controlling strength in a
ramp loading test and time-to-failure in a constant load test are,
indeed, related." This model can also accommodate the effects of
a history of cyclic loading with estimation of the probability of
failure after a given number of cycles for known variability of
the material. Further, it can be fitted to data in which the curve
for stress level vs. time-to-failure under constant load shows a
marked steepening of slope after a period. Such steepening has
been observed with other materials, and is apparently due to
a change in the mechanism controlling fracture. There is some
evidence from long duration tests on wood in bending and shear
parallel to the grain that such a change in slope occurs in wood.
There may be a later flattening of the slope as a threshold stress
is reached and again the model can accommodate a threshold stress.
The curves in Figure 1 illustrate these changes in slope and the
effect of a considerable change in threshold stress is seen to be
negligible on the time-to-failure except near the threshold levels.

Knots and other defects have been shown by Madsen (1971),
Madsen and Barrett (1976), and Spencer (1978) to have an important
influence on the strength of lumber subjected to rate of loading
tests, including ramp loading tests where load is applied in a
succession of steps and held constant for a period at each step.
The lower strength material showed much less decrease in strength

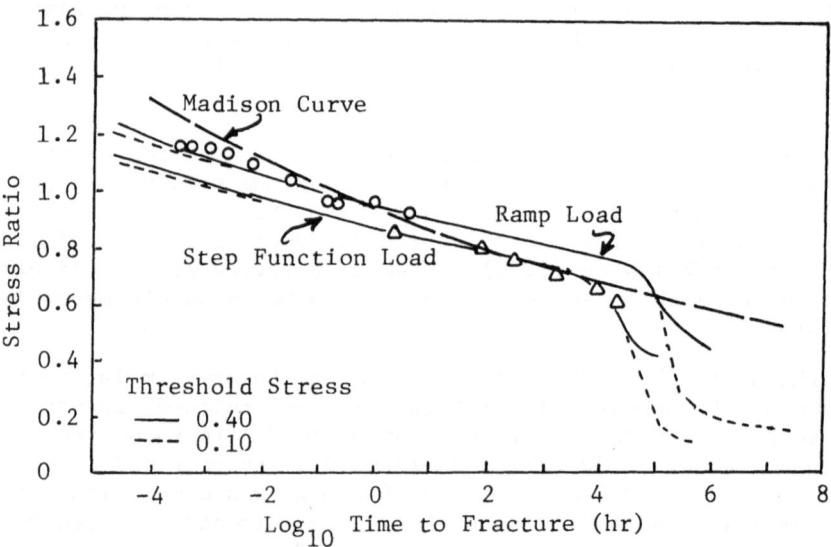

Fig. 1. Effect of assumed threshold stress on fit of damage model
         to Douglas fir data (from Barrett and Foschi 1978).

with time under constant load than higher strength lumber, possibly due to rapid relaxation of stress concentrations at the knots (Madsen 1972). Under rapid rate of loading tests, the low strength lumber even showed a decrease in strength as the rate of loading increased. These results are leading to a reexamination of current methods of allowing for the effects of duration of load in timber design.

Particleboard

McNatt (1975) performed rate of loading tests in bending, tension, edgewise shear and interlaminar shear, and constant load tests in tension, on five types of particleboard representing several species, different types of particles, adhesives, thicknesses and densities. The specimens were conditioned to equilibrium at 73°F and 64% RH.

Bending strength (modulus of rupture) was affected least and interlaminar shear strength was affected most by rate of loading. There was no significant difference between the effects of rate of loading and duration of load on the tensile strength.

McNatt combined his results with those of Lundgren (1969) and with the bending test results of Bryan (1960) and Kufner (1970), $\log_{10} t$ being taken as the dependent variable because time-to-failure depends on the stress level. The resulting relation was given by

$$SL = 83.7 - 8.2 \log_{10} t \qquad\qquad (3)$$

where t = time from start of test (hr).

Lyon and Barnes (1978) obtained similar results in flexural tests on 5/8 inch particleboard for mobile home decking, but urea-bonded boards tended to fail earlier than phenolic-bonded boards at the same stress level.

McNatt and Werren (1976) carried out fatigue tests on three types of particleboard, the minimum stress being 10% of the estimated short-term strength. After 10^7 cycles, the fatigue strengths in tension parallel to the plane of the board and in interlaminar shear i.e. shear parallel to the plane of the board, were 45% and 42%, respectively, of the static strength.

Hardboard

McNatt (1970) reported the results of rate of loading tests in static bending, tension and compression parallel to the surface, interlaminar shear and edgewise shear, constant load tests in tension parallel to the surface and edgewise shear on tempered hardboard conditioned to equilibrium at 75°F and 64% RH. As for particleboard, there were some differences between the regression equations for the various tests. However, they differed only slightly from the following mean regression line for the results of all rate of loading tests.

$$SL = 90.2 - 8.1 \log_{10}t \qquad\qquad (4)$$

The duration of load tests in tension showed almost the same slope of regression line as the rate of loading tests.

The fatigue tests were conducted to maximum stress levels ranging from 37.5% to 90% of the short duration ultimate strength of the control specimens, the minimum stress being 10% of the maximum stress. At 30 million cycles, the fatigue strength in tension was about 40% to 45% of the static strength and in inter-laminar shear was about 40% of the static strength. Specimens which endured 30 million cycles without failure were tested statically and gave ultimate stresses within 10% of the strength of the controls.

Chan (1979) presents a review of the properties of tempered hardboard. He points out that tempered hardboard is very durable for exterior use because the manufacturing process removes much of the sugars and starches on which wood boring insects feed. The equilibrium moisture content is much lower than that of solid wood and so hardboard is not prone to decay even in humid conditions.

DEFORMATION BEHAVIOR OF SOLID WOOD

Creep Under Constant Conditions

Creep occurs in wood at ordinary temperatures, the magnitude of the creep being proportional to the applied stress providing the stress is less than a critical stress equal to p% of the standard test (short time) failing stress, where p is about 50 for bending or compression and 70 to 80 for shear or tension. Irrespective of initial moisture content, green and dry wood exhibit about the same relative creep, i.e. creep expressed as a fraction of the initial (elastic) deformation, when the moisture content and temperature are maintained constant (Armstrong and Kingston 1962).

For stresses exceeding the critical stress, creep increases at an increasing rate with increased stress level so that relative creep is no longer independent of stress level.

Relative creep in bending, compression and tension parallel to the grain are reported by Kingston (1962) to be similar under a given fraction of the short term ultimate stress.

Lofty et al. (1972) reported that the higher the degree of cell-wall crystallinity determined by X-ray diffraction, the less the creep. They also observed that the larger the microfibril angle, the greater the creep.

Moriizumi and Okano (1978) studied the X-ray diffraction pattern of thin wood specimens in tension. Some increase in crystallinity occurred under load but no change was observed in the mean fibril angle. The crystal lattice strain increased during creep, and decreased when the deflection was kept constant.

Relaxation of Stress

If the deformation of a stressed piece of wood is maintained constant, then the stress will diminish with time, a process called "relaxation." Grossman and Kingston (1963) showed that the decrease in stress expressed as a fraction of the initial stress for wood in bending is approximately equal to the relative creep which would be sustained by a similar piece subjected to constant stress, for an initial stress less than the critical stress. Mukudai and Sakamoto (1978) showed a similar result for shear and obtained good estimates of the stress relaxation from a knowledge of the creep compliance of matched specimens. Thus the principle of superposition may be applied, and so the order in which loads are removed or applied does not change the final effect. The deformation under a variety of successive loadings may be calculated from a knowledge of the behavior under a single load applied for a sufficient length of time.

Creep increases markedly with rise in temperature (Davidson 1962). Aoki and Yamada (1977) examined the effect of temperature on the relaxation of hinoki under torsional stress. They concluded that up to about 75°C relaxation appeared to arise from molecular motion in the amorphous regions. At higher temperatures relaxation appeared due to a chemical process, possibly the scission of glucose bonds, the apparent activation energy being 14 kcal/mole compared with 23.7 kcal/mole at the lower temperatures.

Recovery of Deformation

When a load which has been applied for a period is removed, the deformation immediately decreases by an amount equal to the initial elastic deformation. Further recovery occurs with time, indicating that part of the creep was visco-elastic. This recovery is often called the "elastic after-effect." Not all the creep is recovered and the remainder has been called the "permanent set" or "plastic deformation." However, this, too, can be recovered under certain conditions, as will be discussed below.

Influence of Changing Moisture Content

Armstrong and Kingston (1960, 1962) showed that changes in the moisture content of wood under stress produce changes in deformation. Moisture changes also affect the relaxation of stress in wood held to a constant deformation. This phenomenon is called the "mechano-sorptive" effect. Although not strictly a time-dependent effect, except in the sense that the moisture change takes place over some period of time, the mechano-sorptive effect has an important influence on the behavior of wood under service conditions where humidity and temperature often vary widely.

Ellwood (1954) found that beech stressed in tension or compression perpendicular to the grain exhibited creep which increased with increasing moisture content and temperature, and that wood, when drying, showed more creep in compression than in tension. Kingston (1962) also reported that creep in compression parallel to the grain was greater than that in tension under varying moisture conditions. Youngs (1957) observed that higher moisture content and temperature led to increased creep and recovery for red oak stressed perpendicular to the grain in tension or compression. Hearmon and Paton (1964) obtained failure of wooden beams under repeated cycles of moisture change despite the applied stress being only 37% of the short-term strength. Szabo and Ifju (1970) discussed the mechano-sorptive effects in terms of possible changes in the intermolecular bonds.

An excellent summary of the characteristics of the mechano-sorptive effect is given by Grossman (1976). He points out that deformation due to moisture change occurs only with swelling or shrinkage of the cell walls and so occurs above fiber saturation only in species susceptible to collapse. In bending, the first moisture increase leads to increase in deformation, as might be expected, but all subsequent adsorption periods tend to produce some recovery of deformation and desorption leads to increased deformations. The so-called "permanent set" is largely recovered when the unloaded wood is subjected to several cycles of adsorption and desorption. Consequently, any explanation of mechano-sorptive deformation in terms of the breaking and reforming of molecular bonds seems to require the formation of an accompanying elastic mechanism. Because strains at failure are very much higher than those obtained in short-term tests, the failure criterion is unlikely to be one of critical strain.

Arima and Grossman (1978) and Arima (1979) have further
studied the recovery of wood after mechano-sorptive deformation on
a softwood (Pinus radiata) and a collapse-susceptible species
(Eucalyptus regnans) while drying under a variety of regimes of
temperature and relative humidity. For both species, the mechano-
sorptive deformation, i.e. the deformation remaining after removal
of the load and the recovery of the immediate and delayed elastic
deformation, was greatly reduced by a subsequent increase in moisture content and almost eliminated when the wood was finally immersed
in water. This moisture-induced recovery was not influenced by a
delay of two weeks before subjecting the wood to the mositure increase, nor by a temporary increase in moisture content of the wood
before unloading. However, radiata pine specimens subjected to a
temperature of 80°C and Eucalyptus specimens subjected to temperatures of 50°C and 65°C during drying under load lost some of their
capacity to recover. The effect was most pronounced when the
heating was introduced early in the drying process or when it was
coupled with high humidity. Heating above a critical temperature,
i.e. a softening point, causes the formation of a new stable configuration in the wood substance which prevents the deformation from
recovering during humidity cycles. Test results and other information on the softening temperatures of the cell wall components
are given by Hillis and Rozsa (1978), Salmén and Back (1977) and
by Salmén (1979). A well-known practical application of heat is in
the setting of wood to some desired configuration, for example, to
form a bent shape or to produce straight lumber from warped lumber.
Holding the wood to shape under stress during drying at a sufficiently high temperature yields a final product less prone to changes in
shape under subsequent moisture changes.

ADHESIVES AND WOOD-BASED COMPOSITES

Durability of Adhesives

A number of investigators have studied the influence of
exposure outdoors and simulated laboratory regimes on the durability
of adhesives used for plywood and other composite board materials.
All agree that urea resins rapidly lose strength when exposed to
high humidities or water, and when exposed to high temperatures.
The picture, however, is more confused for formulations of exterior-type resins. Gillespie and River (1976) found a wide range
in variability of the wet shear strength of yellow birch and Douglas-
fir plywood panels and shear test specimens after exposure to the
weather for nearly eight years. The panels made with melamine-
urea, hot-press phenol and resorcinol resins all retained 80% or
more of their strength during that period. The small shear test
specimens showed more wood degradation than the panels, with a
consequent greater loss of strength. The melamine-urea resin lost

strength and percentage wood failure more rapidly than the more
durable resins.

Chow and Caster (1978) describe a thermal softening test as a
useful laboratory procedure for ranking adhesives for durability.
Laidlaw (1975) concluded from tests extending over 18 years that
only properly formulated phenolic resins can be recommended for
high hazard exterior conditions, and that excess fillers may cause
poor performance. Steiner and Chow (1975) discussed the effect of
low temperatures on durability of several types of adhesives.

Plywood

The general behavior of plywood under continued stress is
similar to that of solid wood but the lathe checks in the veneer,
the gluelines and the crossed orientation of the plies make some
differences. Kaneda (1979) subjected glue joint shear test specimens of plywood made with several different resins to cyclic wetting
and drying at different temperatures and cyclic changes in relative
humidity at constant temperature. He showed that shear strength
decreased with increased depth of lathe checks. Phenolic resin
adhesives were more durable than the melamine-urea and other resins
used.

Lyon and Schniewind (1978) presented a literature review of
creep studies on plywood and employed linear visco-elastic theory
to develop models for predicting creep in plywood. Their experiments
showed plywood behaved as a linear visco-elastic material for
stress levels as high as 59% of maximum static strength. They
point out that the presence of lathe checks leads to increased
deformation, and that the glue lines and laminating process had
an effect on creep greater than the direct effect of wood compression. However, by determining the constituent properties of the
material from tests on parallel laminated specimens as recommended
by Preston (1954), the creep behavior of plywood was predicted with
good accuracy.

Particleboard and Hardboard

Dinwoodie (1978) has presented a fine summary of the properties of adhesives used in particleboard manufacture and the performance of boards made with them. He points out that so many
parameters affect the performance that changes in any one of them
can result in marked changes in board properties and behavior.
Short duration tests of commercial boards over many years at the
Princes Risborough Laboratory showed that a large measure of uni-
formity of performance existed under dry conditions for the main
types of adhesives, the best overall performance being consistently

given by melamine-fortified urea-formaldehyde resins. However, there was extensive variation in the magnitude of creep among particleboards produced by different manufacturers using the same type of adhesive. For example, both the lowest and highest values of creep occurred with phenol-formaldehyde bonded boards. The density profiles were identical, so the difference in creep performance appears to have been due to resin formulation. Creep deflection was proportional to stress only up to about 20% of the short term failing stress, thereafter increasing at a greater rate with increasing stress level.

Lin and Okuma (1978) observed that the deformation of wet or dry phenolic-bonded boards, subjected to repetitive loading, tended to stabilize but that of boards bonded with urea or of wet boards bonded with melamine-urea progressed until failure occurred.

Unlike creep in solid wood, creep in particleboard and hardboard is greatly influenced by the moisture content of the board. Bryan and Schniewind (1965) obtained relative creep values for particleboard at 12% moisture content which were twice those for the board at 6% moisture content but less than one-third those for the board at 18% moisture content. This was in general agreement with the results reported by Armstrong and Grossman (1972) and shown in Figure 2 for particleboard and hardboard at 6% and 18% moisture content.

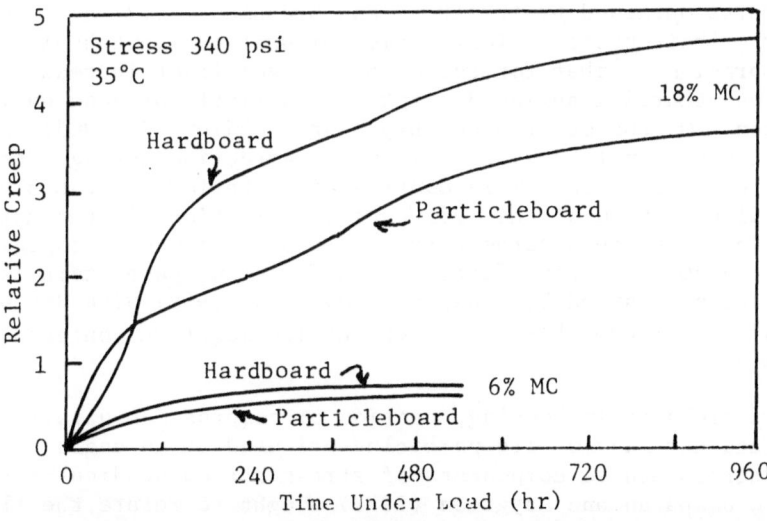

Fig. 2. Effect of moisture content on creep of hardboard and particleboard beams (from Armstrong and Grossman 1972).

Orientation of the strands in particleboard appears to affect the duration of load properties. Hoyle and Adams (1975) compared the effects of duration of load on the flexural strength of commercial oriented strand board, plywood and clear Douglas fir. They concluded that the curves of stress level vs. time-to-failure for the oriented board and plywood were "not measurably different from that for clear wood."

Haygreen et al. (1975) tested urea-bonded and phenolic-bonded commercial particleboard, an oriented flake board and plywood, bending under constant load while being subjected to various constant and cyclic humidity levels. The type of resin had no significant effect on the creep of the particleboards. Flexural creep was directly proportional to stress level for stress levels below 20%. The commercial particleboard showed a sharp increase in creep for relative humidities of 75% and above. Creep increased with increase in relative humidity for the oriented particleboard and for the plywood, but the transition point was less definite. Hall and Haygreen (1978) applied a concentrated load of two years' duration to panels of plywood, a waferboard, a 3-ply oriented strand board and a commercial particleboard. Plywood exhibited the lowest relative creep, closely followed by the waferboard and the oriented strand board, with the ordinary particleboard showing the largest relative creep.

Gressel (1972), Haygreen et al. (1975), Lehmann et al. (1975) and others have shown that changing moisture content has a significant effect on the deformation of particleboard and hardboard, but there have been differences in the results of some investigators. Some results obtained by Armstrong and Grossman (1972) are shown in Figure 3. Their tests indicated similar qualitative behavior during sorption to that for solid wood, significant increases in deflection occurring during desorption and little or none during adsorption. On the other hand, Bryan and Schniewind (1965) and Sauer and Haygreen (1968) obtained more deflection during adsorption than during desorption. This contrast in behavior is probably another illustration of the influence of variations in the materials, formulation or process parameters on the performance of boards. A further example of the effect of manufacturing parameters is the greater creep obtained by Sauer and Haygreen (1968) with dry process than with wet process boards, except at low moisture contents and stress levels.

Particleboard in bending, tension or compression parallel to the surface has many of its particles oriented at an angle to the applied stress and so components of stress act perpendicular to the grain. Gnanaharan and Haygreen (1979) sought to relate the flexural creep behavior of particleboard with that of wood stressed parallel and perpendicular to the grain at relative humidities of 50%, 60%, 70% and 80%. Their randomly oriented waferboard was made from

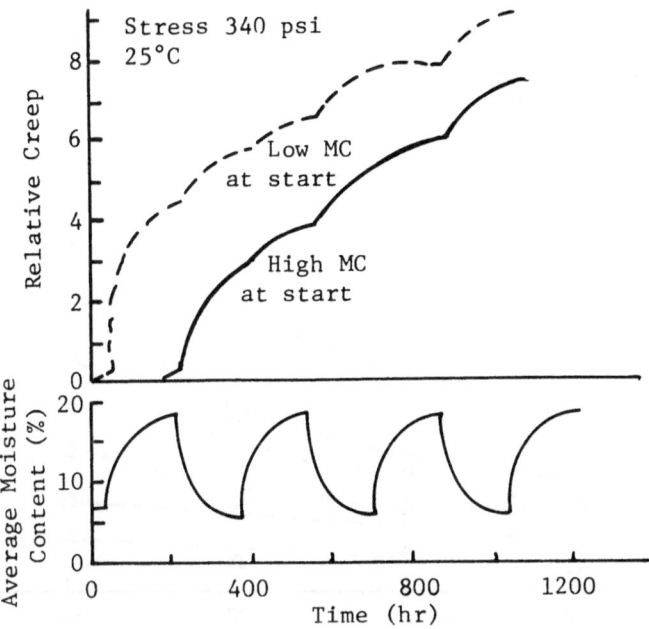

Fig. 3. Effect of initial moisture content and change in moisture content from 6% to 18% on mechano-sorptive deflection of particleboard (from Armstrong and Grossman 1972).

flakes cut from the same tree as that used for the solid wood specimens. The solid wood specimens stressed perpendicular to the grain showed several times the relative creep and recovery of the wood specimens stressed parallel to the grain, the magnitude of the difference increasing with increasing relative humidity. The values for relative creep and recovery of the waferboard were intermediate between those of the solid wood stressed parallel and perpendicular to the grain, the increase in creep for a relative humidity beyond 70% being particularly marked, as may be seen from Figure 4. Somewhat surprisingly, matched solid wood compression and tension specimens stressed perpendicular to the grain did not show the jump in creep at 80% RH observed with the bending specimens stressed perpendicular to the grain.

From work by Bello (1968) and Simpson (1971) it is known that some hardwoods stressed perpendicular to the grain show an increase in moisture content under tension and a decrease in moisture content under compression. The size of the moisture change increases with increased stress and increased initial moisture content. Bello has reported an increase of 1.4% moisture content in wood stressed in compression perpendicular to the grain at 87% RH. Gnanaharan and Haygreen hypothesized that the physical swelling and shrinkage due to stress-induced moisture change, plus the

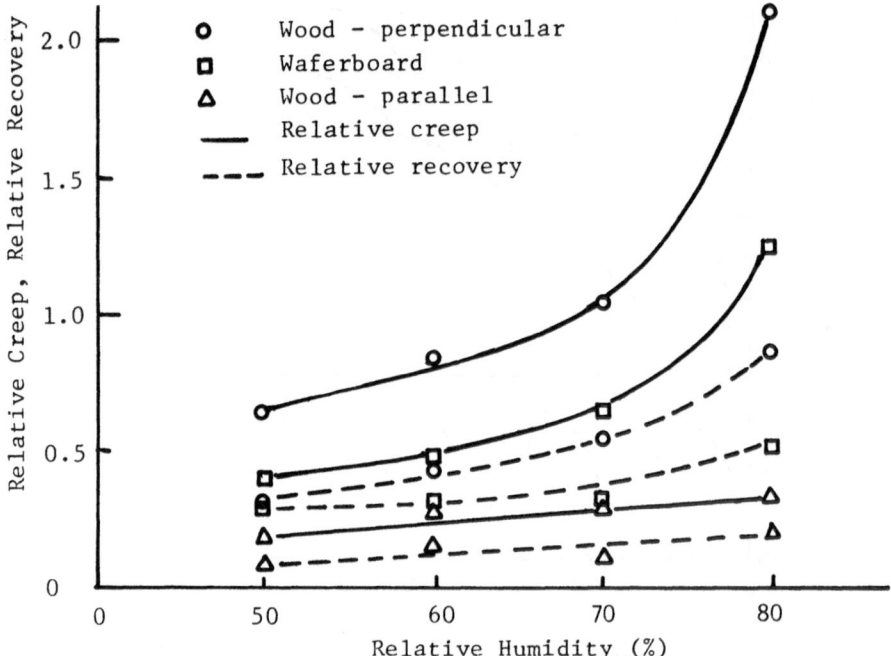

Fig. 4. Effect of relative humidity on relative creep and recovery
         of waferboard and solid wood stressed parallel and per-
         pendicular to the grain (from Gnanaharan and Haygreen
         1979).

mechano-sorptive effect of the change, may be responsible for the
greatly increased creep exhibited when the relative humidity
exceeds about 75%.

     Changes in moisture content produce other effects. In boards
formed under pressure, the adhesive holds the particles in a com-
pressed state. Consequently, particleboard is self-stressed in
compression perpendicular to the grain of the wood. Cycles of
moisture change lead to relaxation of stresses and recovery of the
locked-in deformations with a resulting increase in board dimensions
and some change in moment of inertia. Some of the mechanical
properties are also directly affected but the magnitude depends on
the type of adhesive. Dinwoodie (1978) reported that samples of
urea-formaldehyde bonded boards not under load, when subjected to
50 cycles of humidity change from 30% to 90% RH, retained only 26%,
58% and 45% of their initial tensile perpendicular strength, modulus
of rupture and modulus of elasticity respectively. In contrast,
boards bonded with melamine urea-formaldehyde, phenol formaldehyde
and sulphite liquor retained at least 73% and often more than 90%
of their initial property value. Dinwoodie suggests that the
reduction in properties is not due to any chemical degradation of

the resin but to the effect of mechanical stressing set up by alternate swelling and shrinkage of adjacent particles. This is in line with the reduced creep obtained by Halligan and Schniewind (1972) on pre-steamed particleboard specimens, the steaming treatment causing a lowering of the internal stresses.

Laminated Composites

An enormous variety of composite panels are possible, and many are being made, by bonding faces of different materials to a varied range of core materials. The static properties of such composites can be deduced with reasonable accuracy from the properties of the components. Chow and Hanson (1979) investigated the effect of veneer thickness on the creep behavior of red oak veneered hardboard panels. As might be expected, the thicker the skins, the closer the behavior to that of solid wood. However, little work appears to have been done on the estimation of the duration of load properties of composite panels from a knowledge of that of the components.

REFERENCES

Aoki, T. and T. Yamada, 1977, Chemorheology of wood. I. Stress relaxaton of wood during hydrolysis. Mokuzai Gakkaishi 23(2): 107-113.

Arima, T. and P. V. A. Grossman, 1978, Recovery of wood after mechano-sorptive deformation. J. Inst. Wood Sc. 8(2):47-52.

Arima, T. 1979, Recovery of wood after mechano-sorptive deformation. II. Effects of drying conditions while clamped. Mokuzai Gakkaishi 25(7):469-475.

Armstrong, L. D. and R. S. T. Kingston, 1960. Effect of moisture changes on creep in wood. Nature 185(4716):862.

Armstrong, L. D. and P. V. A. Grossman, 1972, The behavior of particleboard and hardboard beams during moisture cycling. Wood Sc. and Tech. 6:128-137.

Armstrong, L. A. and R. S. T. Kingston, 1968, The effect of moisture content changes on the deformation of wood under stress. Aust. J. Applied Sci. 13(4):257-76.

Armstrong, L. D. and P. V. A. Grossman, 1972, The behavior of particleboard and hardboard beams during moisture cycling. Wood Sc. and Tech. 6:128-137.

Bach, L. 1973, Reiner-Weisenberg's Theory applied to time-dependent facture of wood subjected to various modes of mechanical loading. Wood Sci. 5(3):161-171.

Bach, L. 1975, Failure perpendicular to the grain in wood subjected to sustained bending loads. Wood Sci. 7(4):323-327.

Barrett, J. D. and R. O. Foschi, 1978, Duration of load and
    probability of failure in wood. Part I. Modelling creep
    rupture. Part II. Constant, ramp and cyclic loadings.
    Canadian J. of Civil Eng. 5(4):505-514 and 515-532.
Bello, E. D., 1968, Effect of transverse compression stress on
    equilibrium moisture content of wood. For. Prod. J. 18(2):
    69-76.
Bryan, E. L., 1960, Bending strength of particleboard under long-
    term load. For. Prod. J. 10(4):200-204.
Bryan, E. L. and A. P. Schniewind, 1965, Strength and rheological
    properties of particleboard affected by moisture content and
    sorption. For. Prod. J. 15(4):143-148.
Chan, W. W. L., 1979, Strength properties and structural use of
    tempered hardboard. J. Inst. Wood Sc. 8(4):147-160.
Chow, S. and R. W. Caster, 1978, Relationship of adhesive softening
    temperature to exposure tests for bond durability. For. Prod.
    J. 28(6):38-43.
Chow, P. and R. C. Hanson, 1979, Effects of load level, core
    density and shelling ratio on creep behavior of hardboard
    composites. Wood & Fiber 11 (1):57-65.
Davidson, R. W. 1962, The influence of temperature on creep in
    wood. For. Prod. J. 12(8):377-381.
Dinwoodie, J. M. 1968, Microscopic changes in cell-wall structure
    associated with compression failure. J. Inst. Wood Sci. 4(3)
    :37-53.
Dinwoodie, J. M., 1978, The properties and performance of particle-
    board adhesives. J. Inst. Wood Sci. 8(2):59-68.
Ellwood, E. L., 1954, Properties of American beech in tension and
    compression perpendicular to the grain and their relation to
    drying. Yale University, School of Forestry Bulletin No. 61.
Gerhards, C. C., 1979, Time-related effects of loading on wood
    strength: a linear cumulative damage theory. Wood Sc. 11(3):
    139-144.
Gillespie, R. H. and B. H. River, 1976, Durability of adhesives in
    plywood. For. Prod. J. 20(10):21-25.
Gnanaharan, R. and J. Haygreen, 1979, Comparison of the creep
    behavior of a basswood waferboard to that of solid wood.
    Wood & Fiber 11(3):155-170.
Gressel, P. 1972, The effect of time, climate and loading on the
    bending behavior of wood base materials. Holz als Roh-und
    Werkstoff, 30:259-266, 347-355, and 479-488. (In German).
Grossman, P. U. A. and R. S. T. Kingston, 1963, Some aspects of
    the rheological behavior of wood. III. Tests of linearity.
    Aust. J. Appl. Sc. 14(4):305-317.
Grossman, P. U. A., 1976, Requirements for a model that exhibits
    mechano-sorptive behavior. Wood Sc. and Tech. 10:163-168.
Hall, H. and J. Haygreen., 1978, Flexural creep of 5/8-inch
    particleboard and plywood during two years of concentrated
    loading. For. Prod. J. 28(6):19-22.

Halligan, A. F. and A. P. Schniewind, 1972, Effect of moisture on
    physical and creep properties of particleboard. For. Prod.
    J. 22(4):41-48.
Haygreen, J., H. Hall, K-N. Yang and R. Sawicki, 1975, Studies of
    flexural creep behavior in particleboard under changing
    humidity conditions. Wood and Fiber 7(2):74-90.
Hillis, W. E. and A. N. Rozsa, 1978. Softening temperatures of
    wood. Holzforschung 32(2):68-73.
Hoyle, Jr., R. J. and R. D. Adams, 1975, Load duration factors for
    strand wood, plywood and clear wood. Washington State
    University, Ninth Particleboard Proceedings, p. 83-107.
Hunt, D. G., 1976, Rupture tests of wood chipboard under long-term
    loading. J. Inst. Wood Sc. 7(3):13-21.
Kaneda, H., 1979, Studies on the weatherability of composite wood.
    VIII. The effect of core veneer's lathe check on the dura-
    bility of adhesive joint of plywood under loaded condition.
    Mokuzai Gakkaishi 25(1):43-49.
Keith, C. T., 1971, The anatomy of compression failure in relation
    to creep-inducing stresses. Wood Sci. 4(2):71-82.
Kingston, R. S. T. K. and L. N. Clarke, 1961, Some aspects of the
    rheological behavior of wood. I. The effect of stress with
    particular reference to creep. II. Analysis of creep data by
    reaction-rate and thermodynamic methods. Aust. J. Appl. Sc.
    12(2):211-240.
Kingston, R. S. T. K., 1962, Creep, relaxation and failure of wood.
    Research 15, April 1962:164-170.
Kufner, M., 1970, Creep in wood particleboard under long-term
    bending load. Holz als Roh-und Werkstoff 28(11):429-446.
Laidlaw, R. A. 1975, The durability of glues for plywood manufacture.
    British Research Establishment Information IS9/75:3 pp.
Lehmann, W. F., 1978, Cyclic moisutre conditions and their effect
    on strength and stability of structural flakeboards. For.
    Prod. J. 28(6):23-31.
Lin, T-H. and M. Okuma, 1978. Durability of structural particle-
    board. 2. Some flexural properties in the repetitive loading
    test. Mokuzai Gakkaishi 24(12):879-883.
Lofty, M., M. L. M. El-Osta and R. W. Wellwood, 1972, Short-term
    creep as related to cell-wall crystallinity. Wood and Fiber
    4(3):204-211.
Lyon, D. E. and H. M. Barnes, 1978, Time-dependent properties of
    particleboard in flexure. For. Prod. J. 28(12):28-33.
Lyon, D. E. and A. P. Schniewind, 1978, Prediction of creep in
    plywood. Part I. Prediction models for creep in plywood.
    Wood & Fiber 10(1):28-38.
Lundgren, S. A., 1969, Wood-based panel products as building
    materials. Part I. Swedish Wallboard Association, Stockholm.
McNatt, J. D., 1970, Design stresses for hardboard--effect of rate,
    duration and repeated loading. For. Prod. J. 20(1):53-60.

McNatt, J. D., 1975, Effect of rate of loading and duration of load
    on properties of particleboard. U.S. Dept. of Agric., For.
    Prod. Lab. Research Paper FPL 270.

McNatt, J. D. and F. Werren, 1976, Fatigue properties of three
    particleboards in tension and interlaminar shear. For. Prod.
    J. 26(5):45-48.

Madsen, B. 1972, Duration of load tests for wood in tension perpen-
    dicular to grain. U. of B. C., Dept. of Civ. Eng. Struct.
    Res. Series Report No. 7.

Madsen, B., 1973, Duration of load tests for dry lumber subjected
    to bending. For. Prod. J. 23(2):21-28.

Madsen, B., 1975, Duration of load tests for wet lumber subjected
    to bending. For. Prod. J. 25(2):33-40.

Madsen, B. and J. D. Barrett, 1976, Time-strength relationship for
    lumber. U. of Brit. Columbia, Structural Res. Series Report
    No. 13.

Mindess, S., J. S. Nadeau and J. D. Barrett, 1976, Effect of
    constant deformation rate on strength perpendicular to the
    grain of Douglas fir. Wood Sci. 8(4):262-266.

Moriizumi, S. and T. Okano, 1978, Visco-elasticity and structure of
    wood. IV. Behavior of crystal lattice strain dependent on
    moisture content and time. Mokuzai Gakkaishi 24(1):1-6.

Mukudai, J. and S. Sakamoto, 1978, Evaluating of linear visco-
    elastic behavior of wood. II. On change of load of plate
    shear test specimens being subjected to deflection changing
    with lapse of time. J. Japan Wood Research Society 24(9):
    605-611.

Pearson, R. G., 1972, The effect of duration of load on the bending
    strength of wood. Holzforschung 26(4):153-157.

Preston, S. B., 1954, The effect of synthetic resin adhesives on
    the strength and physical properties of wood veneer laminates.
    Yale University, School of Forestry Bull. No. 60.

Salmén, N. L. 1979, Thermal softening of the components of paper:
    Its effect on mechanical properties. Trans. Tech. Sect., 5
    (3):TR45-50, Canadian Pulp and Paper Association

Salmén, N. L. and E. L. Back. 1977, The influence of water on
    the glass transition temperature of cellulose. Tappi 60(12):
    137-140.

Sauer, D. J. and J. G. Haygreen, 1968, Effects of sorption on
    flexural creep behavior of hardboard. For. Prod. J. 18(10):
    57-63.

Schniewind, A. P. and J. C. Centeno., 1973, Fracture toughness
    and duration of load factor. I. Six principal systems of
    crack propagation and the duration factor for cracks pro-
    pagating parallel to the grain. Wood and Fiber 5(2):152-159.

Schniewind, A. P., 1979, A seven year tale: or how boards have
    breakdown from everyday pressure. Proc. First Inter. Conf. on
    Wood Fracture held at Banff, Alberta, August, 1978. Pub. by
    Forintek Canada Corp., Western Forest Products Laboratory,
    Vancouver, B.C.

Simpson, W. T., 1971, Moisture changes induced in red oak by
    transverse stress. Wood Fiber 3(1):13-21.
Steiner, P. R. and S. Chow, 1975, Low temperature durability
    of common wood adhesives. For. Prod. J. 25(8):26-30.
Spencer, R., 1978, Rate of loading effect in bending for Douglas
    fir lumber. Paper presented at First Wood Fracture Mechanics
    Conference, Banff-Alberta, August 14-18.
Szabo, T. and G. Ifju, 1970, Influence of stress on creep and
    moisture distribution in wooden beams under sorption
    conditions. Wood Sci. 2(3):159-167.
Wood, L. W., 1951, Relation of strength of wood to duration of
    load. U. S. Dept. of Agric., For. Prod. Lab. Report No. 1916.
Young, R. L., 1957, The perpendicular to grain mechanical properties
    of red oak as related to temperature, moisture content
    and time. U. S. Dept. of Agric., For. Prod. Lab. Report No.
    2079.
Ylinen, A. 1957, Zur theorie der Danerstand festigkeit des Holzes.
    Holz als Roh-und Werkstoff. 15(5):213-215.
Gerhards, C.C., 1977, Time-related effects of loads on strength of
    wood. Proc. Conf. on Environmental Degradation of Engineering
    Materials, College of Engineering, Virginia Polytechnic
    Institute, Blacksburg, Virginia.
Hearmon, R. F. S. and J. M. Paton, 1964, Moisture content changes
    and creep of wood. For. Prod. J. 14(8):357-359.

WORKSHOP SESSIONS

WORKSHOP I : MORPHOLOGY AND MECHANICAL PROPERTIES

Co-Chairmen   J. Bodig
                               H. Schreiber

                               J. Balatinecz
                               M. Bariska
                               A. Bolton
                               A. DiBenedetto
                               B. Harris
                               B. Hope
                               B. Kubat
                               R. Landel
                               J. Manson
                               R. Mark
                               G. Matolscy
                               L. Nicolais
                               M. Piggott
                               E. Roffael
                               Z. Rigbi

I. INTRODUCTION

The workshop was charged with the responsibility to identify
the specific issues relating to mechanical and morphological
characteristics of adhesion and those of bonded cellulosic and
wood composites allowing for their increased use. The group
consisted of representatives from the fields of composite
materials, polymer physics and engineering, cellulose science and
wood technology. The discussion indicated a lack of
communication among these sectors to the detriment of the
subject. In view of the enormous potential which all categories
of composites hold for growth and broader application, the
workshop served an important purpose simply in providing a forum
for the exchange of pertinent information. It is the opinion of
this group that similar exchanges on closely identified topics
should be fostered at regular intervals.

The workshop deliberated technical challenges to be met in
opening new horizons for wood and cellulose based composites.
These deliberations also heeded the matter of effectively using
available resources which may not, in fact, be unlimited in
quantity or quality. An individual viewpoint to this end may be
found as part of the proceedings of this symposium. The workshop
group approved, as a majority, the inclusion of the viewpoint as
being pertinent to its work without necessarily endorsing it.
The following were identified as steps to follow toward
developing the recommended research needs:

- identification of the scope of discussion

- selection of broad areas for further discussion

- detailed discussion of specific relevant scientific and
  technical issues

- recommendations for research

II.   SCOPE OF DISCUSSION

Broadly, research on adhesion is viewed as a route to
optimizing material efficiency and economics of cellulosic and
wood composites. In turn, this optimization demands detailed
fundamental knowledge of the mechanical properties of wood,
cellulosic composites and their adhesives. These composites can
be classified into two major areas:

(a)   Conventional cellulosic and wood composites, in which the
        matrix occupies a relatively small proportion of the
        total volume.

(b)   Non-conventional composites, in which the matrix occupies
        a relatively large proportion of the total volume.

In addition to the relative volumes of matrix and constituent

the balance of mechanical properties in composites may be viewed
from several different vantage points. Thus, it is important to
understand not only the mechanical properties of the finished
composite but also the properties of matrix and other components
individually prior, and during the bonding process.

Information therefore is needed on issues such as damage done
to the surface and interior of the components during their
preparation; characteristics of the adhesive prior to bonding;
their physical, mechanical and especially rheological
characteristics during bonding and those of the final product.
It is felt that only with the above specifics well understood,
will it be feasible to predict the performance of composites
under specified use conditions.

Many composites based on synthetic polymers may be considered
to be two-phase systems; for these there has evolved a broad base
of fundamental information interrelating composition, morphology
and mechanical properties. Cellulosic and wood composites on the
other hand contain significant void volumes, moisture, etc and a
multi-phase system must therefore be considered. In this
category of composites phenomenological studies have prevailed,
and it is not clear to what extent the fundamentals applying to
the former group of materials may be applied to describe the
behavior of the latter. Clearly, a key objective of future
studies should be to evolve an analogous fundamental basis for
the interpretation of mechanical properties in the wood- and
cellulose-based composites and for the development of improved
composites.

The following properties, characteristic of wood and/or
cellulosic components, were prominent in setting our research
targets:

- inherent variability of wood, particularly in its morphological
  and mechanical characteristics.

- orthotropic character

- hygroscopic properties

- unusual depth of an interface consisting of damaged material
  bonded by adhesive

- limited bio-, thermal - and chemical stability

- low energy requirement in processing

- attractive thermal/electrical insulation properties

- renewability of the primary resource

III.   MAJOR AREAS OF CONCERN

The workgroup identified the following major areas of concern for detailed analysis:

(a)   Exploitation of the special characteristics of conventional cellulosic and wood composites and their mechanical properties.

(b)   Morphology as it relates to the mechanical properties of components and bond strength.

(c)   Identification and weighting of the variables affecting the mechanical properties of the composites.

(d)   Flaw identification and detection techniques.

(e)   Stresses in composites generated through external loads and environmental variables.

(f)   Applicability of existing theories to cellulosic and wood composites.

(g)   More information on mechanical properties to develop high performance composites.

(h)   Variability of components and end-products.

(i)   Mechanical properties of adhesives during bonding and in the finished product.

(j)   Visco-elasticity of adhesives and composites.

IV.   SUBSTANCE OF DISCUSSIONS

(a)   Conventional Composites

Limitations to the performance of conventional cellulosic and wood composites such as paper, fiberboard, particle-board, "comply", plywood, parallel laminated veneer, and glued laminated timber, were considered. Among the leading limitations the following were identified:

- high cost of adhesives,

- low end-joint strength

- hygroscopic swelling

- low thermal conductivity of the wood (relating to long cure times)

- lack of fundamental understanding of adhesive-component interaction.

Offsetting these limitations, the following were cited as advantages of conventional composites:

- high strength and stiffness/weight ratios

- low energy requirements in processing

- wide range of shapes and forms

- greater load - carrying efficiency than unmodified lumber.

Research recommendations to remedy the limitations without compromising advantages included the following:

- improvement of final end-joint strength.

- use of fracture mechanics to analyze, understand and predict failures in the adhesive and glue line.

- correlation of manufacturing variables with mechanical properties of the structure.

- improved stiffness through the use of high modulus reinforcing components.

- reduction in the amount of adhesive without detriment to property balance.

- fundamental understanding of interaction between adhesive and wood/cellulose (component), to interpret bond formation and resulting mechanical characteristics.

(b)   Morphology

There was broad agreement that a fundamental understanding of the morphology of composites was a prerequisite to an interpretation of mechanical properties and to a prediction of the use characteristics of these structures. The current state of knowledge was discussed at length and the following recommendations for research were formulated:

- Development of improved quantitative analyses and measurement techniques, such as electron and optical microscopy, x-ray analysis, improved microtome methods.

- Application of statistical methods to describe morphological variations.

- Better treatment of fiber orientation,distribution and fiber mass distribution problems, experimentally through digitizer and image analyzer methods, and theoretically through better mathematical modelling.

- Identification of the effects on mechanical properties of

morphological elements such as tracheids, ray cells; included also is the effect of inherent flaws and of those created in processing, on mechanical property balance. Clarification of this issue may be speeded by the use of model systems in which chemical and physical flaws are deliberately incorporated.

- Development and increased use of non-destructive test methods to clarify the behavior of composites in elastic (Hookean) range, and to correlate this behavior with morphological detail.

(c)   Visco-Elastic Properties

There was unanimous agreement that cellulosic and wood composites are visco-elastic (V-E) in nature. Research in this area should be strongly emphasized, because the knowledge gained will be essential to interpreting the origin and mechanisms of failure, of excessive deformation, of dynamic and time-dependent responses, cyclic load and environmental responses of structures. V-E studies are called for in the overall composite and equally, visco-elestic descriptors of the adhesion process are required. The following major categories of endeavor were recommended;

- Assemble pertinent V-E data for typical categories of composites. As a first step, a survey of world literature should be considered.

- Evolve a relevant interpretative and predictive model for V-E behavior.

- Perform experimental tests to check the model, and refine in the light of results.

- Perform sensitivity studies with the refined model to ensure its useful response to variables such as:

  · cyclic loading

  · cyclic environmental changes

  · chemical and physical effects of use

  · wide range of load frequencies

- Work toward evaluation of standardized measurement techniques for V-E property evaualtion

- Establish applicability of existing V-E theories (developed for relatively simple single or 2-component systems) to the present multi-component cases

- Define V-E property parameters for unsteady state conditions, including aspects of the adhesion process itself (e.g. glue penetration, cure, etc) and during variations in temperature and moisture-content.

The results of studies noted above should lead to a variety of benefits, including:

- increased performance reliability under field conditions

- greater sophistication in meeting performance demands through more precise selection of components

- more efficient use of raw materials

- extension of the usable range of physical and environmental conditions for these composites.

(d)   Environmental Stresses:

It was the concensus that stresses produced in composites, primarily by variations in the temperature and moisture level, often limit their broader use. Little is known presently about the exact mechanisms whereby these environmental variables exercise their effect. It was therefore recommended that the following studies be undertaken:

- Establish the effect of moisture and temperature on such factors as fiber swelling, fiber strength, etc.

- Establish moisture distribution patterns and moisture migration during steady and unsteady state conditions

- The role of environmental cycling on the life expectancy of the composite should be established; specifically, applicability of metallurgical theories and analyses should be considered

- Micro-mechanical and thermodynamic disciplines should be applied to the study of the dissimilar effects of environmental factors on the stresses existing in the adhesive/adherent parts of the composite.

(e)   Other Areas of Concern:

This section summarizes less complete discussions of additional factors germane to the workshop objectives:

- Forecasting long-term mechanical behavior through rapid test procedures was a repeatedly expressed need. Non-destructive and dynamic evaluations appear to be feasible approaches. The ability to forecast long-term behavior accurately implies also an understanding of ageing processes in the composite.

- Much theoretical and empirical expertise in the overall area of material composites is now available. Various questions with research implications arose from this fact. Thus, the applicability of theory developed for two-phase composites should be tested using available data on the more complex, multi-phased wood and cellulose composites. A critical evaluation of processes developed for existing conventional composites may provide guidelines for the formulation and production of newer generations of wood and cellulosic composites.

- Opportunities exist for the use of high strength and high stiffness components of wood fibers in new high-performance composites. To realize these opportunities, studies of the properties of cell and cell-wall components are called for. The self-bonding capabilities of natural wood should be characterized quantitatively. Further challenges were recognized in the relative flammability and bio-degradability of wood. Suggested approaches include the use of silaceous materials (man-made petrifaction effects). The use of wood components as fillers and reinforcing agents in thermoplastics was also cited as a research target.

- It was noted that non-wood plant materials represented valuable sources for composite formulations. Little systematic knowledge now exists on the behavior of such composites. Basic research on the mechanical properties of this class of materials is strongly recommended to further their development and use.

V. CONCLUSIONS

The deliberations of this workshop group attained their general objectives of focussing attention on the importance of mechanical and morphological characteristics in the evolution of wood-and cellulosic-based composites.

While the weight of reporting on various aspects of these discussions may seem uneven, it should be recognized that no such weighting applies to the importance of the various topics involved. Rather, this state of affairs reflects the constraints inherent in a small group and a limited time-frame. Indeed, several of the subjects discussed - e.g. morphological characterization, visco-elastic behavior, were considered to be of such complexity and importance, that a recommendation was made to convene workshops on these more specifically defined topics, in order to formulate more cogent research strategies.

ADDENDUM TO GROUP I REPORT (R.E. Mark)*

We were told on Tuesday noon that our purpose was to concentrate on areas where the use of cellulosic materials be expanded particularly where these materials occupy the larger (by volume and/or weight) part of composite products.

*An expression of private views by the author.

The premise is that cellulosic materials generally, and wood and wood fibers in particular, are relatively abundant and available to perform many new tasks. The statement was made to us that cellulosic material is "cheap" and "plentiful". I question this premise.

While there may be a few areas of the world where wood is cheap and plentiful, the world's wood producing capability is in considerable jeopardy[1]. Some of the major pressures on forest production capability already are:

1. Population growth - Most of the countries that possess high growth rate forests -- the tropical countries -- have population growth rates that will double populations in 20 to 25 years. This population growth puts great pressure on forest lands, which are being cleared for conversion to food production and other uses at the rate of at least 80,000 square miles per year in those countries.

2. Per capita forest products consumption increases. Two major factors are accounting for increased use of wood-based materials per capita. The first is the worldwide explosion in communications, records-keeping and packaging, all of which require vast increases in paper production. The second, very recent, trend is for widespread conversion of small energy systems (such as home-heating) to utilize wood fuel. In large areas of the temperate zone countries, traditional wood-using industries are finding it hard to obtain needed raw material. because of the competition with wood-burners. For example, sawmills and paper mills in some countries are facing chronic shortages due to the demand for fuel wood, and pulp "logs" with top diameters as small as 1.5cm are being trucked to pulp mills.

3. Forest growth rates are declining in many areas. Principally, the causes of reduced forest growth rates are environmental in nature. The specific assaults on the forest vary with the area of the world, but some examples are:

   a.   Acid rain and snowfall, caused by high atmospheric transport of sulfur-containing anthropogenic gases (which react with water vapor to form sulphuric acid), and photochemical production of nitric acid over long distances through progressive reactions of industrially produced ozone with atmospheric nitrogen. The areas most affected at present are (a) Northern Europe, and (b) Northeastern North America. Forest growth rate declines of approximately 10% in Scandinavia have been attributed to acid precipitation, according to studies done in Norway, Sweden and other countries.

[1] Editor's note: see "The World's Tropical Forests - A Policy, Strategy, and Program for the United States", Dept. of State Publication 9117 (1980).

b. Industrial air pollution, caused by anthropogenic gases.
    Examples abound from all industrialized countries, or
    areas with large numbers of automotive vehicles.

c. Erosion. In almost all parts of the world, erosion
    losses today are several times as great, perhaps 10 or
    more times as great, as soil replacement is occurring
    through natural weathering processes. The principal
    cause is the vast acceleration in land clearing and
    agricultural (cropland) conversion. This is a very
    severe problem in Latin America.

d. Soil depletion. Most nutrients esential for plant growth
    are incorporated into the tissues of trees as they grow.
    Removal of the wood, and especially bark and foilage,
    results in permnent removal of these nutrients from the
    growing forest. Removal of forest growth can not be
    sustained indefinitely without replacement of all the
    elements that are in short supply in a given geographical
    area. How soon growth rates start to fall off noticeably
    depends on the particulr element(s) that are critically
    needed by the tree, and the original abundance of such
    element(s) in the soil. A noticeable problem in several
    European countries now, requiring expensive applications
    of fertilizers, usually be helicopters.

e. Soil compaction. A very widespread phenomenon,
    especially in high alumina content soils in the tropics.
    Cutting of forests on the (extensive) soils of this type
    in countries like Burma and Thailand and Brazil results
    in the rapid formation of highly impermeable soil that
    has little viability for sustaining new forest growth.
    The process is called laterization.

f. Desert encroachment. Lands of marginal rainfall that lie
    between deserts and forested areas have, in recent years,
    been subject to overgrazing and plowing that has greatly
    reduced the moisture-holding ability of the soil. The
    result is an enlargement of the area of desert at the
    expense of brushland and forest land. The problem has
    become severe in areas of India and Pakistan, and
    critical for several countries of sub-Saharan Africa.
    New dust bowls have also been created in the western
    hemisphere that have impacted upon forest growth.

The result of all these and other pressures is that the world's
temperate zone forests have declined moderately, and the world's
tropical zone forests have declined precipitously. The National
Academy of Sciences of the United States has completed a study
showing that the world's potential tropical forest area of 6.4
million square miles had been cut to an estimated 3.74 million
square miles by 1975. If the rates of forest destruction were
constant, it would take but 50 years to eliminate the world's
tropical forests altogether, but the study concludes that "the
rates are not constant -- they are accelerating rapidly".

Despite this grim outlook, there is a widespread tendency to
look upon forests as our next great source of both energy and
chemical feedstocks. Although there are current ecomonic
slowdowns in some places (e.g. in the U.S. housing industry), the
world wide demand for forest products is very great, and many
existing industries cannot now fill their needs for wood fiber
adequately. They are understandably apprehensive about plans for
medium-to-large size wood-burning power plants that each require
an area of 50-mile radius to be dedicated exclusively to
supplying the plant at the center with wood.

We must always examine, with scientific skepticism, the
projections of resource availability that the optimists claim are
out there waiting to be exploited. It is true that "forests are
renewable", but the statement holds true only if the productive
land base is not diminished beyond the capability of better tree
and plant fiber strains and improved land management to make up
for the loss, and if the losses in soil productivity and
atmospheric quality are reversed. The major trends today suggest
that nothing of the sort is liable to happen soon.

REBUTTAL TO WORKSHOP I ADDENDUM (B.W. Burgess)

I think these are valuable cautions which must be kept in mind,
but at the same time we must realize that there are a number of
other factors which must be taken into account.

1. No one is suggesting that we plan for harvests beyond the
   capacity that the land can provide and some of the workshop
   reports recognize this. Capacity varies greatly from country
   to country and from region to region. There are parts of the
   world where substantially more cellulosic material could be
   obtained than is now being taken from the land.

2. Cellulosic materials include plants as well as trees and we
   know that annual crops, such as kenaf, may be used to
   supplement trees as a raw material for pulp, with a much higher
   yield of fibre per acre.

3. It is well known that forest yield can in many, many locations
   be greatly increased by proper forest management.
   Increasingly, this is being recognized by governments and
   industry alike and both are collaborating more and more on
   better forest management. It should also be remembered that
   the annual deposition of foliage provides a significant
   portion of the humus and nutrients required for tree growth.

4. The "explosion in communications" referred to in the addendum
   will not be dependent on paper alone, but on electronics as
   well. This is already evident with the introduction of the
   Videotex home information systems now appearing in many
   countries.

The extent to which the forest is relied upon to supply
society's varied needs will depend on its ability to provide,

<u>under proper management</u>, the cellulosic materials required. It will also depend on many economic factors which vary from country to country and which take into account the cost of cellulose-based materials, relative to the cost of other alternatives and the economics that prevail.

WORKSHOP II: SURFACE AND INTERFACE

Co-Chairmen	J. Oliver
	J. Schultz
Secretary	D. Smith
Members	E. Andrews
	J. Bristow
	W. Hillis
	W. Johns
	K. Keirstead
	L. Sharpe
	P. Stenius
	J. Wellons

I. INTRODUCTION

The first session was taken up with a general discussion of bonding problems in wood based composites as related to the surface and interface variables. The list of keywords (distributed prior to this meeting) served as a framework for the discussion.

The state of knowledge concerning surface wetting of wood by adhesives and penetration to form bonds was reviewed by the wood-experts in the group. The application of classical techniques to these phenomena was discussed and it was evident that to elucidate these problems more information on the kinetics of wetting and surface penetration under practical joint-forming conditions was needed.

The importance of glue penetration and thickness of the glue layer to interfacial bond strength was discussed. It was apparent that a better definition of the wood surface and of the characteristics of an adhesive which affect physico-chemical interactions between the two phases, was required for a better understanding of bond fracture. The use of natural glues based on bark extractives was considered; at present this was also limited by the need for improved knowledge of glue wetting, spreading, and penetration characteristics.

Innovative approaches to adhesive utilization and application were considered in the light of the economics of wood composite production. The use of dry coating techniques and of liquid glues was outlined and the need for minimizing the amount of glue emphasized. It was pointed out that whereas cellulosic-filled polymers utilize large amounts of polymer, and veneers and plywood utilize continuous glue layers, particle-board requires up to 12 per cent adhesive and flake-board 2 per cent. The group concluded that the latter products present the greatest difficulties from an analytical point of view and that products having a major amount of polymeric adhesive were of lower priority in the present discussion. In view of the fact that future forest management techniques would probably result in shorter growing times and smaller trees it was decided to orient the discussion to surface phenomena related particularly to flake- and particle-board and similar materials. Comparable surface interaction would occur in other products such as plywood and veneered boards.

The problems underlying optimum bonding between these wood surfaces and the glue were discussed in detail. It was evident that the surface of wood was poorly understood and the contribution of physical and chemical effects to bonding were ill-defined. The need for a model for wood surfaces to allow analysis of surface energetics and analysis of bond energy failures was emphasised in view of the difficulty of applying existing surface concepts.

II.   SURFACE/INTERFACE RELATIONSHIPS

After a general introductory discussion concerning the nature of the problem, the working group decided to consider the surface/interface relationships at a number of different levels which may be summarized as follows:

Molecular

- the level at which molecular interactions are relevant and where, in particular, the reactions between different molecular functional groups need to be considered.

Micro

- the level at which chemical interactions may be considered between different phases and where the distribution of different chemical species and surface energetics are relevant.

Mini

- the level at which the physical properties, porosity and roughness, of the solid phase need to be considered in the context of how these properties affect the bond formation.

Macro

- the level at which the geometrical configuration of the particles of the solid phase determine the extent of particle-particle contact.

Mechanical

- the level at which the mechanical properties, elasticity etc. of the different phases and the interphase created at the boundary between them influence the properties of the bond.

Maintenance

- the level at which temporal effects related to the degradation of the component phases as a result of diffusion, oxidation, hydrolysis etc. affect the bond permanence.

At each of these levels, the group considered the state of present knowledge, what needs to be known to extend the state of knowledge and what modifications might be introduced to increase the strength, permanence and ease of formation of the final cohesive bond.

The major problems identified at each size level and possible techniques for their solution are listed in Table 1. The variables which affect each situation are also categorized. This analysis demonstrated a need for the application of more sophisticated and innovative physical and chemical techniques to

TABLE 1

Surface and Interface Problems in Wood-Based Composites

Problem	Techniques Applicable	Variables
a) Molecular Level - up to 1nm		
1. Identify chemical species on natural wood surface	Fourier Transform IR, Laser Raman, and x-ray photo-electron (ESCA) spectroscopy, atomic labelling	temperature time Surface formation — A i) chemical ii) physical species
2. Determine nature of interfacial bonds with adhesive	Fracture energy analyses, chemical blocking, model compounds	A Adhesive System - B
3. Determine nature of improvements in interfacial bonding	(improvements needed)	
b) Micro Level - up to 10nm		
1. Identify distribution of chemical species	histological, electron microscopy	A
2. Characterize surface energetics	contact angle (advancing, receding), calorimetry, adsorption studies (improvements needed)	A

TABLE 1 (cont'd)

Surface and Interface Problems in Wood-Based Composites

Problem	Techniques Applicable	Variables
3. Characterize micropore structure	gel exclusion, gas diffusion	A
4. Evaluate surface chemical modification	oxidation, reduction, blocking, grafting, coupling, glow-discharge, thermo-mechanical	A
5. Evaluate surface physical modification	extraction, dehydration, solvent	A
6. Examine effects of modification of surface content of wood particles	photo-acoustic, torsional braid, T_g of bound resin	B
c) Mini Level - up to 100μ m		
1. Characterize effect of topography	Scanning electron microscopy (SEM)/ photogrammetry, replication	A
2. Characterize effect of topography on interphase, viz. i) spreading ii) penetration and its control iii) dimensional change iv) bond strength v) pressure or absence of adhesive	SEM of sections, staining of components, porosimetry, gel permeation, chromatography of adhesive systems, liquid tensiometry (improvements needed)	A & B

TABLE 1 (cont'd)

Surface and Interface Problems in Wood-Based Composites

Problems	Techniques Applicable	Variables
d) Macro Level		
1. Examine particle – particle contact (distribution of adhesive, voids)	serial sections, micro-encapsulated tracers, x-ray diffraction analysis of probe particles	A particle size distribution, surface area, orientation, C
2. Examine penetration of adhesive and localisation	as 1 (improvements needed)	A + B + C
e) Mechanical Level		
1. Determine properties of wood and adhesive phases	conventional	species, B
2. Determine properties of interphase	impregnation of thin wood films, interphase bonding, non-destructive testing	A + B
f) Maintenance of Properties		
1. Characterisation of locus of failure	microscopic	A + B, environmental history: water diffusion, biological hydrolysis, oxidation, plasticisation, embrittlement, fatigue, ultra violet radiation
2. Characterisation of bond strength	conventional fracture mechanics	as 1

TABLE 2

SUMMARY OF SELECTED TOPICS RELATED TO THE SURFACE AND INTERPHASE

CHEMICAL ASPECTS

          Chemical species on wood surface
          Nature of interfacial bond
          Chemical modification of surfaces

PHYSICAL ASPECTS

          Characterization of micropores and capillaries

          Characterization and influence of topography and adhesive
          properties on interphase formation

          Physical modification

TECHNOLOGICAL ASPECTS

          Particle-particle contact
          Localising adhesive

FAILURE ANALYSIS

          Effect of environmental history on:

                    locus of failure
                    joint strength

study the properties of wood and adhesive surfaces and also the "interphase" formed between them. The latter region of interaction is, at present, little understood.

The research approaches were then reviewed as to their importance in understanding and improving bonding in current wood-based composites.

The ideal interphase would require, at least the following characteristics:

- a minimum of low cost adhesive
- adequate mechanical properties
- strong wood-adhesive interface
- durability
- ease of fabrication
- dimension stability

III.   RESEARCH RECOMMENDATIONS

In the light of these considerations the research problems of greatest priority which appeared to be most amenable for solution were selected for future work (see Table 2).

WORKSHOP III: DURABILITY

Co-Chairman   R. Gillespie
                                    R. Pearson

                                    E. Back
                                    A. Challis
                                    J. Dinwoodie
                                    E. Graminski
                                    J. Koutski
                                    R. Kreibich
                                    A. Lambuth
                                    E. Plueddemann
                                    H. Sazaki
                                    P. Steiner
                                    J. Talbot

I. INTRODUCTION

Due to lack of time, the group concentrated its attention on cellulosic-based composites in which the resin occupied a small proportion of the volume and served primarily as the bonding agent. Other systems exist and require study; some mention of them is made in this report.

A useful distinction was noted between the terms "durability" and "permanence". Durability was defined as the resistance to degradation under the influence of stress and environmental conditions. Permanance was defined as the resistance to degradation due to age only, i.e. where stress is absent or negligible. Although relevant to paper, permanence is not applicable to composites of the type considered by the group. These composites, even if not subjected to external stress, are in a state of self-stress near the bonded regions owing to their formation by a pressing operation.

II.   SOURCES OF CELLULOSIC MATERIALS

The main sources of cellulosic materials which might be tapped for composites are as under:-

1. Agricultural residues such as corn cobs, rice hulls (which are also a fine source of silicon), and peanut hulls.

2. Waste paper, especially from municipal garbage (e.g. the techniques for removing plastics from the paper have now been developed and one-way shipping pallets have been made from waste paper pulped and formed to shape in a mould with phenolic resin as the adhesive, but the main problem with the use of some types of waste is that the fibers are short and the delignification tends to reduce the stability of the product.)

3. Forest residues.

4. Milling and manufacturing residues.

5. Species and trees not presently suitable for solid products.

III.   CONSERVATION OF FOSSIL FUELS BY DEVELOPMENT OF NEW   ADHESIVES

Strong efforts should be made to develop adhesives from non-petroleum sources. Some waste products, such as black liquor from kraft paper mills and biomass from non-commercial trees and plants, are potential sources of phenolic and other resins. Unfortified tannin adhesives have been developed in South America to the commercial stage.

Cement as a binder is being increasingly used in Europe. Wood-cement panels are proving suitable for external cladding and have potential for houses in the Third World.

Improvements in melamine, urea and polyvinyl acetate resins should be sought to upgrade their performance in the more severe environments.

Isocyanates, acrylics and other resins offer promise for further development of cellulosic composites.

Foamed plastics need to be developed to extend adhesive use and to reduce damage due to high compaction of the substrate.

Until excellent substitutes for phenolics become available the use of the very small amount of petroleum needed for making phenolics may be of more national benefit than its use for energy.

IV.   DURABILITY CLASSES

There is an urgent need for:

a)   The standardization on an international basis of test procedures and equipment for evaluating the durability of cellulosic composites in relation to end-use requirements,

b)   The definition of end-use requirements on the basis of factors affecting durability, and

c)   The development of durability classes to cover a range of end-uses and products.

More than the two durability classes of "exterior" and "interior" are highly desirable for the economical use of materials, some of which are in short supply in various parts of the world. It is recognized that additional classes would introduce some complication and extra cost in production, marketing and maintenance of inventories.

The specification of expensive waterproof adhesives for materials which may be subjected to one, or at most a few, short periods of moderately hazardous conditions, e.g. rain at a building site, not only increases costs but inhibits the development and use of alternative satisfactory adhesives.

Four durability classes are suggested with good durability against the hazards outlined for the respective classes.

Class 1 - Exterior: Wetting and drying cycles, exposure to UV radiation, large temperature variations, attack by micro-organisms.

Class 2 - Protected exterior: Cycles of wide range of relative humidity and temperature changes.

Class 3 - Interior, humid conditions: Cycles of high and low
        humidity.

Class 4 - Interior, dry conditions: Cycles of moderate changes
        in relative humidity.

In addition, the composites should have adequate mechanical
properties and limited creep under the respective environmental
conditions.

In setting requirements for these classes, absolute values for
properties retained after aging, or accelerated aging, treatments
should be specified rather than percentage retentions.

V. EXTENT OF EXISTING KNOWLEDGE ON DURABILITY TESTS

Present test methods enable the prediction of the suitability
of wood composites made with current adhesives for traditional
applications. They also provide the background information that
explains why different adhesives perform well, or poorly, in
particular service environments. All laboratory short and long
term tests, as well as weathering tests and historical
performance, show that phenolic and resorcinol adhesives are
stronger and more durable than the wood itself.

Accelerated laboratory aging tests are capable of ranking
adhesives with regards to their resistance to one or more
degrading influences. These tests are not sufficiently
selective, however, to enable an adhesive less durable than wood
to be classified as adequate for a particular service situation.

VI.   BASIC RESEARCH NEEDS

Fundamental research should be conducted to obtain a coherent
picture of the properties and performance of adhesives and
composites made from them. This should be done both for the
initial state and after a period, or periods, of subjection to
stress and changing environmental conditions.

Topics for study include the following:

1.   Mechanism of the degradation of wood.

2.   Identification of parameters affecting

      a)   optimization of adhesive properties

      b)   optimization of the adhesive and cellulosic
            substrate when combined in some practical
            proportions

      c)   reproducible manufacture of the composite.

3.   Mechanism of failure of the adhesive-substrate interface.

4.   New techniques to measure specific properties of
      interfaces, fracture energy and long term performance
      properties of composites to enable composites to be
      classified into durability classes.

5.   Rheology of cellulosic composites, including load,
      moisture and temperature cycling before and after aging
      treatments.

6.   Effect of moisture and temperature on the bond between
      adhesive and cellulosic interface and on the cohesive
      properties of the adhesive, including the influence of
      stresses induced by changes in moisture and temperature.

7.   Stability of the interface against hydrolysis, oxidation,
      differential movement, hysteresis and plasticization by
      moisture.

8.   Effect of fiber morphology on performance of composites
      where the cellulosic component acts as reinforcement of
      the resin matrix. Morphological properties of importance
      include length to width ratio, fiber length distribution,
      cell wall thickness, density, fiber bundles, fibrillation
      and three-dimensional orientation.

9.   The development of models for predicting the long term
      performance of cellulosic composites in a variety of
      service environments.

10.   Chemical characterization of components and effects on
      durability.

11.   Fire resistance and flammability of the composites.

WORKSHOP IV: FABRICATION, APPLICATION AND ECONOMICS

Co-Chairman   T. Maloney
                       M. O'Halloran

Secretary   G. Woodson

                       B. Burgess
                       S. Chow
                       A. Geller
                       B. Hope
                       S. Mathur
                       G. Mavel
                       K. Shen
                       M. Yan

I. INTRODUCTION

The ultimate aim of this meeting is to make possible the wider
use of low cost cellulosic or wood-based composites so as to
conserve resources. The overall objective is to identify the
technological factors which limit the use of cellulosic- or wood-
based materials in composites.

The limitation of time called for a concentration on large-
volume uses, if a significant impact is to be made, on expanding
the use of the composites under consideration. The large volume
use selected is in the area of shelter materials or systems and
the work group elected to concentrate on this area. It is
recognized, however, that small-volume and specialized composites
have their place and may in time use significant amounts of
cellulosic or wood materials.

For the purpose of the work group a composite was defined as
any wood or cellulosic material requiring an adhesive or some
type of bond or self bond. The spectrum of materials thus runs
from glued homes, through laminated members, glued components
including trusses, wall sections, I-beams, end and edge glued
lumber, plywood, comply, particleboard, fiberboard, mouldings, to
materials having a high level of matrix bonding material and a
low wood content. The wide use of such materials is presently
possible in the "wood using" countries because of their building
practices; however, improvements in the composites under
consideration may well enhance the use of such materials in
countries not now using wood heavily.

It was recognized that the best basic philosophy for forest
products use at present is to keep the raw material as large as
possible to conserve production energy and adhesive. The major
new uses of available raw material, however, would require the
production of many types of composites. Raw material
considerations were considered as part of integrated forest
products operations except where agricultural wastes are used.

Development of new composite materials, or expansion of present
production will have to be based on trees or other biomass of
decreasing quality. This includes tree tops, bark, stumps and
roots, agricultural waste, and municipal waste. Competition for
most mill waste is already heavy from present board plants, pulp
mills, and energy generation.

The forest products industry--particularly that portion based
on comminuted wood--has progressed to the stage where additional
fundamental research is now needed if significant gains are to be
made. It has been proven that many different products are
possible. However the solid wood products industry appears to
have a technical position similar to the pulp and paper industry
of some 75 years ago---in need of fundamental reserch to make
significant new advances. If the analogy of the pulp and paper
industry is appropriate, then major technical advances are
possible for the solid wood products industry.

The work group divided the discussion into six major subject
categories and addressed each with due consideration given to
current trends. Both basic and applied research problems were
identified, Manufacturing and end-use (final application) were
covered in such a way, to avoid when possible, the efforts of the
other work groups.

The six subject categories were:

1)   economics
2)   raw material
3)   energy
4)   adhesives/bonding systems
5)   product/processing
6)   other.

II.   ECONOMICS

General Statement - The group's opinion was that research into
specific economic matters did not seem fruitful when
consideration is given to the rapidly changing political climate.
Economic matters, when appropriate, were considered with each
research consideration. Specific economic considerations would
have been possible when considering narrow product lines.
However, under the constraints of looking at large-volume uses,
relatively low cost composites are the most important as they
will be mostly directed towards shelter products.

III.   RAW MATERIAL

General Statement - The group saw that the raw material source
could range from the standing forest to agricultural residues
and/or the municipal waste of large cities. Collection and
subsequent utilization of each type will have distinct problems
and product opportunities. It is therefore recommended that a
systems approach be utilized to develop an analysis of the
availability and usefulness of the various possible raw
materials, giving due consideration to impact on related systems.
It is also recognized that in many places surpluses of wood or
cellulose may not exist, therefore, economic use of the raw
material should be of paramount concern.

Specific Considerations

1. A systems approach should be developed to review the technical
   possibilities of utilizing all parts of the tree and forest.
   The forester/ecologist must determine the impact of removing
   material from each forest type. The technologist must develop
   a best-use product for each portion of the forest removed so
   that the technology is available when the economic opportunity
   is at hand.

2. The composition of the standing forest should be characterized
   as to species, amount of material, and type (e.g. tops, limbs,

roots, leaves, etc). Further, mechanical, physical and chemical properties should be determined. This characterization of raw material should be expanded to agricultural products and by-products, potential industrial wastes. An inventory of raw material available for use will thus also be available for production planning.

3. Agricultural and other waste products should be evaluated as to their cost of recovery.

4. Resource availibility should be reviewed by region with consideration given to reforestation of likely areas--especially in countries with a small or marginal forest base.

5. Systems should be developed to clean dirty raw material.

IV.   ENERGY

General Statement - Energy is clearly a serious consideration for the forseeable future with regard to both manufacturing technique, end-use utilization and the energy saving potential of the product itself. Efforts should be directed toward reducing the energy consumption of the process. Shelter is a basic need for people and care is needed in determining whether suitable raw material should be diverted to energy generation.

Specific Considerations

1. Research should analytically reassess the major production processes with emphasis on producing similar products with new or innovative energy efficient methods. The energy cost of pollution control should receive special attention.

2. Water removal is the major energy-consuming process in the production of wood-based materials. Basic research is desperately needed to review the physics of water removal from wood products and its effect on gluability.

3. Product development is suggested for panels and/or building systems with improved thermal insulation characteristics.

4. Continuation of the energy analysis of construction systems is recommended. Emphasis should be given to both components and the building's indoor environment as related to energy saving construction. Problems with indoor pollution must be identified and solutions to the problems provided.

5. Consideration should be given to the use of wood as an energy source, however the production of longer-life products for shelter is considered of greater need. Wood should not be directed toward energy simply because of its easy availibility. Its place in terms of energy should be established by a systems approach.

V. ADHESIVE/BONDING SYSTEMS

General Statement   Non-petroleum based adhesive systems have
been given considerable attention. It is, however, the group's
opinion that benzene feedstock will be available over the long-
term from both petroleum and coal resources, despite short term
shortages. The desirability of producing adhesive systems which
are independent of oil is however acknowledged. Most important
is the development of a better understanding of the fundamentals
of bonding in wood and cellulosic materials. Both bonding and
degradation of the bond require a better understanding. The
development of systems with improved bonding should also
contribute to improved dimensional stabilization.

Specific Considerations

1. Systems which have a low cost base should receive special
   attention. This would include sulfur, bark, waste liquor from
   pulping, and carbohydrate-based adhesives from wood or other
   agricultural wastes from pulping. Also included are chemical
   modifications to the wood to enhance present bonding systems or
   provide self-bonding.

2. Adhesive characteristics for production, and field construction
   were reviewed. Efforts should be directed toward such systems.

   a)   Production systems:

   -    bonding at high moisture content has high priority with
        regard to energy saving and improved product performance.
        Less water would be removed from the cellulosic or wood
        raw material to allow bonding as it is now practiced.
        Further, water wouldn't have to be added to the products
        after bonding the composite to equilibrate them to
        expected ambient conditions.

   -    theoretical advances in basic understanding of bonding is
        of special concern.

   -    modification of the substrate to optimize properties
        through oxidative coupling or other such approaches is
        potentially of great value.

   -    other, more obvious characteristics, include fast cure
        times, long shelf and pot-life, low cost, and exterior
        performance.

   -    emphasis should be placed on dimensional stability in
        combination with strong, durable bonds.

   b)   Field applied systems:

   -    Major advances in field construction systems are seen if
        field-applied, cold setting, gap-filling, and rigid

adhesives, could be developed. This not only would open
the way for improved shelter systems but would conserve
material because sophisticated engineering designs can be
implemented which use less material in constructing
buildings. As much as 50% of wood could be saved in
constructing a typical home.

VI.   PROCESSING/PRODUCT

General Statement - Processes and resulting products cover a
wide variety of potential products. Shelter-oriented materials
may be well served by performance-based standards which tend to
encourage the product technologist. Determination of any hazards
from raw materials on final products relative to health and
product liability prior to commercilization must be done. The
entire production process must be reconsidered based on
fundamental research to provide a thorough understanding of the
process.

Specific Comments

1.   Recommend the continued development of performance or
        end-use standards. The group believes that such
        definitive standards would become targets for development
        of new products to meet those standards.

2.   Combinations of products for shelter, or development of
        components to improve construction systems, is
        encouraged.

3.   For composite matrerials, improvements in processing and
        materials produced should be possible by understanding
        the fundamentals of manufacturing, particularly in the
        areas of fiber generation, drying, adhesive application,
        and pressing.

4.   Basic research is needed to minimize water pollution
        problems in the wet process board industry. Solving
        these problems opens the way for expansion of this
        industry.

5.   Health considerations in both the manufacturing and
        product use requires extra effort. Utilization of
        chemicals which may exhibit instability should be
        carefully examined before they are used in products.
        Appropriate evaluation methods need to be developed.

6.   Air in energy-tight shelters needs to be investigated
        with respect to indoor pollutants. Methods to clean
        indoor air such as filters and heat exchangers are
        required. The entire building should be considered as a
        system.

7.   Development of moulding techniques at less than 20

atmospheres is of importance in furthering the expansion
of products. Fundamental work in process limitations,
plasticizing, particle size, and chemical treatments for
further moulding should be investigated.

8.   Fire retardant and preservative treatment for decay and
         termite resistance is encouraged. Such wood products
         have much longer lives in use; thus, there is a reduced
         demand on the forest for raw material for the manufacture
         of replacement building materials.

VII.   OTHER CONSIDERATIONS

The working committee suggests that the utilization of wood-
based and cellulosic materials has such impact on world economics
that the formation of a continuing special NATO project program
should be implemented. Such a project would provide
international interchange of ideas with continued emphasis on
related disiplines to broaden the approach. Of particular
importance would be the continued interaction of those in
attendance at this meeting so as to build upon the strong new
inter-relationships developed during the session.

VIII.   AREAS OF OPPORTUNITIES

Better understanding of the fundamentals of bonding and
processing in cellulosic and wood-based composites will lead to
wider use of such materials. The wider use of such materials is
extremely important as they are based on renewable resources.
The potential exists for using renewable resources for all or
part of the bonding system. Thus, non-renewable resources can be
conserved for uses which cannot draw upon alternate raw matrials
as a base. Understanding of bonding phenomena will provide the
opportunity to conserve material in construction. Reliable
bonding systems can lead to sophisticated engineered structural
systems requiring much less material.

Successful research should expand the use of cellulosic and
wood-based composites by increased raw material utilization and
providing a further saving by using more efficient building
construction systems.

IX.   RESEARCH PRIORITY RECOMMENDATIONS

Numerous research opportunities are noted in the main body of
this meeting report. As with any such endeavor, a few
opportunities must be given the highest priority for greatest
impact upon the problems to be solved. The highest priorities by
general consensus as recognized by the Working Group chairmen
are:

1. Following a systems approach, develop an inventory of available
   forest, agricultural, and municipal waste materials suitable
   for use in cellulosic and wood composites. This inventory is

to include pertinent physical and chemical properties of the various materials. It is emphasized that availability of the raw material is of great importance in this inventory since removal of any material has to be balanced in the ecosystem involved.

2. Bonding in cellulosic and wood-based composites needs to be fully understood. Improvements in bonding systems can then be based on this fundamental research. A very important new adhesive system will be one for gluing under moist conditions, (about 18% moisture content and relatively low temperatures). This includes both plant manufacturing of products and field applications. In many products, bonding and dimensional stabilization should be considered as one, because the latter property can be as important as bonding itself.

3. Energy analyses of construction systems should be continued and expanded. Conservation of energy through the use of present and new composites, in components, are of great importance. The impact of energy conservation on building interior environments needs to be assessed completely. Indoor pollution problems need to be solved and techniques developed to assess environmental impacts from all building materials either used now or in the future.

X. SPECIAL RECOMMENDATIONS

Continuation of the activities of the Materials Science Panel/NATO Science Committee, is strongly endorsed in order to encourage scientists from several disciplines to meet on the subject of adhesion in cellulosic and wood-based composites.

SUMMARY OF CONFERENCE

The Conference was planned by an Organizing Committee at the Pulp and Paper Research Institute of Canada (PAPRICAN), Pointe Claire, Quebec, with Dr. S. Chow, Canadian Forest Products Ltd., Vancouver, British Columbia as chairman and Dr. P. Lepoutre, PAPRICAN, as secretary. A group of 55 invited specialists in wood, paper and polymer science, representing 10 countries attended the Conference.

The <u>Conference objectives</u> were stated in the opening address by Mr. B.W. Burgess, President of PAPRICAN. Two main types of wood and cellulose composites were seen:

- Cellulosic fibres and wood materials used in limited amounts as reinforcement and filler in a synthetic polymer.

- Plant materials used as the main component (of the order 80-90%) in a composite with a synthetic or native polymer added as modifier or adhesive.

The purpose of the Conference was to promote development of a broader use of cellulose, wood and other plant materials which occur as an abundant and renewable resource in many parts of the world. The Conference program focussed on the cellulose-polymer interaction in composites especially in relation to adhesion problems, with plant material being the main component. Also included in the program were synthetic polymers, e.g. plastics reinforced with cellulosic fibre. In conclusion Mr. Burgess asked two key questions: What is unique about cellulosics and how can their properties be best utilized in composite materials?

In the first part of the program, three halfday sessions with three plenary lectures in each session dealt with <u>the components</u>, <u>adhesion</u> and the <u>critical properties of the composites</u>, respectively.

The first plenary speaker, Dr. R.E. Mark, College of Environmental Science and Forestry, SUNY, Syracuse, New York, outlined the present knowledge of <u>"Molecular & Cell Wall Structure of Wood"</u>. He particularly stressed the complex morphology and physical structure of wood tissue and the intricate distribution of chemical components in cell walls and middle lamellae of typical wood species which are pertinent problems in the formation of composite materials.

Dr. R.E. Kreibich, Weyerhaeuser Technology Centre, Tacoma, Washington, described <u>"Structural Wood Adhesives - Today and Tomorrow"</u>. This is a story of success for the petroleum - based synthetic adhesives used for bonding wood-based structural building materials from wood components. In many tests, e.g. tensile strength, mechanical and chemical durability, the adhesive bonds were found to be superior to the wood components used. The recommendations were to develop resins based on renewable raw materials, with fast curing times at ambient low temperatures, to secure resources and save energy.

Dr. P.R. Steiner, Forintek Canada Corp., Western Forest Products Lanboratory, Vancouver, British Columbia, took notice of Dr. Kreibich's recommendations in his lecture on <u>"The Forests as a Source of Natural Adhesives"</u>. He discussed the use of lignin, bark and foliage as raw materials for adhesives. Lignin is an inexpensive polymeric phenol which has promise for large-scale utilization. The use of bark materials for adhesives is still largely exploratory with limited commercial success only in Australia and South Africa. Tree foliage is an abundant resource, comprising some 10% of mature trees and up to 25% of young trees. Addition of foliage materials to cellulosic composites bonded with phenolic resins, has been shown to decrease unwanted release of formaldehyde from board and laminates.

The second plenary speaker, Dr. L.H. Sharpe, Bell Laboratories, Murray Hill, New Jersey, started the session on "Adhesion and Composites" with a lecture on <u>"Interaction at Interfaces"</u>. He particularly stressed that the thin surface layer of a material (a few molecules thick) has a physical and chemical structure which is different from its bulk. The surface layers usually absorb extraneous substances (even inert gases), may be crosslinked by radiation (visible light or UV) or by bombardment with inert gas molecules, and in the formation of composites may react with the adhesives and other additives.

Dr. A. DiBenedetto, University of Connecticut, Storrs, Connecticut, described in his lecture methods for <u>"Evaluation of Fiber Adhesion in Composites"</u>. The adhesion between fibers and bonding resins or polymer matrices was shown to be very dependent on fiber surface treatments, viz. coupling agents (e.g. silanes), sizing agents (e.g. PVA to prevent fiber-to-fiber contact) and lubricants (e.g. amines to decrease fiber breakage during fabrication and use). The reinforcing effects of fibers are functions of their length-to-diameter ratio.

Dr. J.D. Wellons, Oregon State University, Corvallis, Oregon, analyzed the various contributions to <u>"Bonding in Wood</u>

Composites". These included mechanical interlocking, physical
attraction (secondary bonding forces), and chemical bonding.
Different phases in the process of adhesive bond formation were
recognized: wetting of fiber surfaces, resin penetration into
cell lumen and voids, formation of chemical bonds, and
conditioning of the composite formed (water migration by flow and
diffusion, etc.). Also adhesive-less bonding of surface-treated
wood components was considered.

The third Plenary lecture dealt with critical properties of
wood-based composites and their determination. Dr. R.F. Landel,
Jet Propulsion Laboratory, California Institute of Technology,
Pasadena, California, discussed "Problems Encountered with
Conventional Fibre-Reinforced Composites". For polymeric
materials and composites the mechanical behavior above the glass
transition temperature (T_g) is well understood and can be
interpreted, based on molecular parameters. Below T_g there are
no adequate experiments available and no molecular theory has
been developed. Physical ageing, strongly time-dependent
properties and complex failure behaviour, are other problems
relevant to composite materials.

Dr. R.H. Gillespie, Forest Products Laboratory, Madison,
Wisconsin, presented the various types of "Wood Composites" and
also outlined their appearance and physical properties. The wood
elements may be fibers, particles, flakes, wafers, strands,
veneers or sawn lumber board. They may be combined with other
elements like plastic films or foams, metal foils or mineral
particles, using synthetic or native resins as adhesives.
Mechanical properties, water swelling, resistance to fungi and
insects, emission of formaldehyde, durability in weathering, and
dimensional stability, are important characteristics of wood
composites.

The "Time-Dependent Properties" of wood-based composites were
treated in a concluding lecture by Dr. Ronald G. Pearson, North
Carolina State University, Raleigh, North Carolina. Pure woods
have been thoroughly studied in creep experiments with loads of
long duration (ramp loading to failure, step loading, and
constant loading). The failing stress (SL) vs. time (t) is well
interpreted by the exponential "Madison equation". Logarithmic
time relations were reported for failing stresses of particle
board and hardboard in bending, tension and shear mode stress.
Moisture content was found to be a critical parameter for creep
and recovery under load, both for pure wood and wood-based
composites.

The plenary sessions included discussions of the material
presented in the lectures. It was pointed out that information
of suitable fabrication methods, bonding mechanisms, mechanical
properties, effects of moisture, durability and weathering
resistance, etc. for cellulosic and wood-based composites are
incomplete and scattered. The wood structure itself has a very
complex morphology and an intricate chemical composition.
Analytical data on chemical composition of surfaces in fractured

wood have only recently become available by ESCA and similar methods. Non-destructive dynamic-mechanical measurements, which have proven so successful for synthetic polymer systems, have only recently been applied to wood-based composites.

The question was raised by Dr. R.E. Mark if wood really would be available as an inexpensive ("cheap") and abundant ("plentiful") resource in the future. Dr. Mark pointed out that the world's wood producing capability is in considerable jeopardy, mainly due to population growth, decreasing forest area, increasing consumption of forest products per capita, and declining forest growth rates in many areas. This last factor was attributed to acid rain and snowfall, other industrial and urban air pollution, erosion of land, soil depletion, soil compaction, and desert encroachment. While forests have declined moderately in the world's temperate zones, the tropical forests have declined precipitously in recent years. In spite of these negative factors the forests are now looked upon as the next great resource of energy (by wood burning) and chemical raw material (forest waste and biomass).

Despite these cautionary remarks the meeting still projected a wider use of inexpensive forest products in composites for construction materials. However Dr. Mark's comments were approved for inclusion as an addendum to the agenda together with a rebuttal by Mr. B.W. Burgess which addressed four main points:

· increased forest growth and useful yields through improved forest management

· prevention of soil depletion by depositon of annual foliage

· impact of electronic communication on paper demand

· affect of economic factors on alternate uses of forest products.

The last half of the meeting was organized as a workshop with the participating experts divided into four groups to discuss research and development programs in specified areas.

Group I with Prof. H.P. Schreiber, Ecole Polytechnique, Montreal, Quebec, and Prof. J. Bodig, Colorado State University, Fort Collins, Colorado, as Co-Chairmen, worked on "Morphology and Mechanical Properties". Some characteristic properties of wood and cellulosic components - both favourable and unfavourable - were found important for setting research targets on composites: inherent variability of wood and fibers, orthotropic character, hygroscopic properties, unusual depth of interface consisting of damaged material to be bonded by adhesive, limited bio-, thermal and chemical stability, low energy requirements in processing, attractive thermal and electrical insulation properties, and renewability of the primary resource.

Limitations of the performance of conventional cellulosic and wood-based composites were indicated: high cost of adhesives, low end-joint strength, hygroscopic swelling, low thermal conductivity of the wood (giving long cure times) and lack of fundamental understanding of adhesive-component interactions.

Among advantages of composites offsetting these limitations were: high strength and stiffness/weight ratios, low energy requirements in processing, wide range of shapes and forms, and greater load-carrying efficiency than un-modified lumber.

The following research recommendations were made on <u>morphology</u> of wood-based composites: improved quantitative analyses, application of statistical methods to describe variations, better treatment of fiber orientation and fiber mass distributions, identification of effects on mechanical properties of morphological elements such as fiber cells, tracheids and ray cells and increased use of non-destructive test methods to clarify the elastic (Hookean) behaviour of composites and to correlate this behaviour with morphological detail.

Research on the <u>visco-elastic</u> (V-E) <u>properties</u> of cellulosic and wood-based composites was strongly emphasized: assembling pertinent data for various types of composites and evolve a predictive model for V-E behaviour related to composition, curing, moisture content, etc., and to variables such as cyclic loading, cyclic environmental changes, chemical and physical effects of use, and a wide range of load frequencies. The resulting data should lead to increased performance reliability of the composites under field conditions, greater ability to meet performance demands, more efficient use of raw materials, and extension of the usable range of composites.

<u>Environmental effects</u> on wood-based composites often limit their broader use. The following studies were recommended: establish the effect of moisture and temperture on swelling, modulus, strength, etc., the moisture distribution pattern and moisture migration during steady and non-steady state conditions, the role of environmental-cycling on the life expectancy of the composite, and micromechanical studies of environmental factors on stresses in adhesive/adherent parts of the composite.

Group II with Dr. J.F. Oliver, Xerox Research Centre of Canada, Mississauga, Ontario, and Dr. J. Schultz, Universite de Haute-Alsace, Mulhouse, France, as Co-Chairmen, worked on "<u>Surface and Interphase</u>". The group defined the bonding problems in wood-based composites as related to surface properties, interface variables, surface wetting and penetration by the adhesive. The need for systems requiring minimal amounts of adhesives of low price was emphasized, e.g. particle-board requiring up to 12 per cent adhesive and flake-board only 2 per cent. The surface of wood was found to be poorly understood and the contributions of physical and chemical effects to bonding ill-defined. Interface interactions in composite bonding were analyzed as phenomena at succeeding levels of magnitude from molecular interaction to mechanical properties of bulk materials. The ideal interphase would require a minimum of adhesive, giving strong wood-adhesive bonding with adequate mechanical properties, dimensional stability, durability and also ease of fabrication (i.e. low temperature, short time).

Research problems of greatest priority were selected, e.g. on the molecular level advanced surface study techniques (FTIR, ESCA, labelling and staining, etc.) were recommended. On levels of larger magnitude, more conventional techniques could be applied. In failure analysis it was considered important to know the environmental history and its effects on joint strength and locus of failure.

Group III with Dr. R. Gillespie, Forest Products Laboratory, Madison, Wisconsin, and Dr. R.G. Pearson, North Carolina State University, Raleigh, North Carolina, as Co-Chairmen, worked on "Durability", primarily of cellulose-based composites with small amounts of resin as the bonding agent. A distinction was noted between "durability", which is defined as resistance to degradation under stress (mechanical, environmental, etc.), and "permanence", as resistance to degradation due only to age. Only "durability" was considered in this case. An inventory of possible sources of cellulosic materials for composites listed agricultural residues, waste paper, forest residues, wood-milling residues, and trees not presently suitable for solid products.

Strong efforts were recommended to develop new adhesives to save fossil fuels, e.g. by using non-petroleum waste products (lignin, tannin, native resins, etc.), cement, improved melamine, urea and polyvinylesters for severe environments, and foamed plastics (to prevent compaction). The group found an urgent need to specify and standardize durability classes of composites related to end uses, e.g. for exposed exterior, protected exterior, humid interior, and dry interior conditions, respectively. Accelerated durability tests were found adequate for phenolic and resorcinol resin bonded composites which are shown to be more durable than the wood itself. Accelerated laboratory ageing tests are, however, not sufficiently selective to reveal whether an adhesive less durable than wood, is adequate for a particular service situation.

The basic research needs in this area should be aimed to give a coherent picture of properties and performance of adhesives and composites made from them, e.g. include the following topics: mechanism of wood degradation, optimization of adhesive and adhesive/cellulosic substrate properties and of reproducible manufacture of composites, mechanism of failure of the adhesive/substrate interface, stability of an interface against repeated water swelling, hydrolysis, oxidation, differential movement,and hysteresis and plasticization by moisture.

Group IV with Dr. T.M. Maloney, Washington State University, Pullman, Washington and Dr. M.R. O'Halloran, American Plywood

Association, Tacoma, Washington, as Co-Chairmen, worked on
"Fabrication, Application and Economics". The group found that
the ultimate aim of the meeting was to make possible a wider use
of low cost cellulosic or wood-based composites as construction
materials so as to conserve scarce or potentially-scarce
resources. Discussions and recommendtions were focussed on
manufacturing and end-use and were divided into six major subject
categories.

Specific research into economics did not seem feasible but
economic matters were considered, when appropriate, with each
research approach discussed.

Raw materials were seen as ranging from standing-forests to
municipal waste. The group recommended a systems approach to
analyze the availability and usefulness of the various raw
materials, e.g. how to use all parts of the tree and the forest,
how to evaluate the use of agricultural and waste products, and
how to increase the resources available, e.g. by reforestation
and better management in forestry and agriculture.

Energy was rated as a serious matter in manufacturing and
application techniques and in the energy-saving potential of the
product itself. Specific considerations were given to the need
to reassess the production process, introduce new and innovative
methods, and pay attention to the cost of pollution control in
fabrication. The physics of water removal from wood products
should be studied as a basis for energy-economy in production.
Development of insulating board with improved energy-conservation
properties in application, was recommended, taking into
consideration whole building construction, and problems with
indoor air pollution. The use of wood as a raw material in
composites for building construction was considered a greater
need than use of wood as an energy source.

For adhesive bonding systems non-petroleum products (e.g. bark,
sulfur, waste liquor, wood carbohydrates, and agricultural waste)
were regarded as having important potential use. The group's
opinion, however, was that because benzene feedstock was so
valuable for bonding-resin production, it use will not be
displaced over the long term. High priority should be given to
the development of adhesives suitable for low-energy production
field construction, bonding at high moisture content, fast cure
and long shelf and pot-life.

In the process/product area it was important to cover a wide
variety of potential markets, specified by performance-based
standards to encourage product development. Improvements in
processing and materials production should be possible by
understanding the fundamentals of manufacturing in the areas of
fiber generation, drying, adhesive application and pressing
(curing).

Health considerations, both in manufacturing and product use,
require extra effort, e.g. in the use of chemicals which exhibit

instability and in fire retardant and preservative treatments.

   In conclusion the utilization of cellulosic- and wood-based
materials was found to have sufficient impact on world economics
to warrant formation of a <u>continuing special NATO
project/program</u>.

Prof. Bengt Ranby
Dept. of Polymer Technology
The Royal Institute of Technology
S-100 44 STOCKHOLM, Sweden

LIST OF PARTICIPANTS

Dr. E.H. Andrews
Department of Materials
Queen Mary College
Mile End Road
London E1 4NS
England

Dr. E.L. Back
Fibre Building Board Dept.
Swedish Forest Products Lab.
Box 5604
Drottning, Kristinas VAG 55
S-11486 Stockholm
Sweden

Prof. John Balatinecz
Faculty of Forestry
University of Toronto
Toronto, Ontario M5S 1A1
Canada

Prof. Michael Bariska
Swiss Federal Institute of Technology
Dept. of Forestry and Wood Research
Binzstrasse 39
8045 Zurich
Switzerland

Prof. J. Bodig
Dept. of Forestry and Wood Science
Colorado State University
Fort Collins, Colorado 80523
U.S.A.

Dr. A.J. Bolton
Dept. of Forestry and Wood Science
University College of North Wales
Bangor, Gwynedd LL57 2UW
United Kingdom

Dr. J.A. Bristow
Swedish Forest Products Lab.
Box 5604
Drottning, Kristinas VAG 55
S-11486 Stockholm
Sweden

Mr. B.W. Burgess
Pulp and Paper Research Institute
of Canada
570 St. John's Blvd.
Pointe Claire, Quebec H9R 3J9
Canada

Dr. A. Challis
Science Research Council
P.O. Box 271
3-5 Charing Cross Road
London WC2H OHW
England

Dr. S. Chow
Manager, Research and Applied Science
Wood Products Manufacturing Division
Canadian Forest Products Ltd.
15th Floor, 505 Burrard Street
Vancouver, B.C. V7X 1B5
Canada

Dr. A. DiBenedetto
Vice-Principal, Graduate Studies
University of Connecticut
Storrs, Conn. 06268
U.S.A.

Dr. J.M. Dinwoodie
Princess Risborough Laboratory
Princes Risborough, Aylesbury
Buckinghamshire HP17 9PX
England

Dr. A. Geller
Institut für Papierfabrikation
Alexanderstrasse 22
D-61000 Darmstadt
West Germany

Dr. R. Gillespie
U.S. Forest Products Laboratory
P.O. Box 5130
Madison, Wisconsin 53705
U.S.A.

Dr. E.L. Graminski
National Bureau of Standards
A-309 Polymers Bldg.
Washington, D.C. 20234
U.S.A.

Prof. Brian Harris
Dept. of Materials Science
Bath University
Claverton Down
Bath BA2 7AY
England

Dr. W.E. Hillis
C.S.I.R.O.
Dept. of Building Research
P.O. Box 56
Highett, Victoria 3190
Australia

Prof. Brian B. Hope
Dept. of Civil Engineering
Queen's University
Kingston, Ontario K7L 3N6
Canada

Dr. Wm. E. Johns
Washington State University
Wood Technology Section
Pullman, Washington 99164
U.S.A.

Dr. K.F. Keirstead
Royal Military College
Kingston, Ontario
Canada

Dr. J.A. Koutsky
Dept. of Chemical Engineering
University of Wisconsin
3016 Engineering Bldg.
Madison, Wisconsin 53706
U.S.A.

Dr. R.E. Kreibich
Chemistry Department
Weyerhaeuser Technology Centre
Panther Lake
Tacoma, Washington 98401
U.S.A.

Dr. B. Kubat
Chalmers University of Technology
Dept. of Polymeric Materials
FACK S-402
20 Gothenburg
Sweden

Dr. Alan Lambuth
Boise Cascade Company
P.O. Box 50
Boise, Idaho 83728
U.S.A.

Dr. R.F. Landel
Jet Propulsion Laboratory
California Institute of Technology
Pasadena, California 91125
U.S.A.

Dr. Pierre Lepoutre
Pulp and Paper Research Inst. of Canada
570 St. John's Blvd.
Pointe Claire, Quebec H9R 3J9
Canada

Dr. T.M. Maloney
Dept. of Materials Science & Engineering
Washington State University
Pullman, Washington 99163
U.S.A.

Dr. John A. Manson
Materials Research Centre
Building 32
Lehigh University
Bethlehem, Pa. 18015
U.S.A.

Dr. R.E. Mark
Dept. of Wood Science and Technology
S.U.N.Y.
Syracuse, N.Y. 13210
U.S.A.

Dr. V.N. Mathur
Canadian Forestry Service
Dept. of the Environment
Ottawa, Ontario K1A OH3
Canada

Dr. G.A. Matolcsy
Ontario Research Foundation
Dept. of Applied Chemistry
Sheridan Park
Mississauga, Ontario
Canada

Dr. G. Mavel
IRCHA
18, bis. boul. de la Bastille
75012 Paris
France

Dr. Luigi Nicolais
University of Naples
Instituto di Principi di Ingegneria Chimica
Piazzale Techio
80125 Napoli
Italy

Dr. M.R. O'Halloran
American Plywood Association
7011 - 19th Street
Tacoma, Washington 98466
U.S.A.

Dr. John F. Oliver
Xerox Research Centre of Canada
2480 Dunwin Drive
Mississauga, Ontario L5L 1J9
Canada

Dr. R.G. Pearson
Dept. of Wood and Paper Science
North Carolina State University
Raleigh, N.C. 27607
U.S.A.

Dr. M.R. Piggott
Dept. of Chemical Engineering and
Applied Chemistry
University of Toronto
Toronto, Ontario M5S 1A1
Canada

Dr. E.P. Plueddemann
Dow Corning Corporation
Sagina Road
Midland Michigan 48640
U.S.A.

Prof. Bengt Ranby
Dept. of Polymer Technology
The Royal Technical College
100 44 Stockholm
Sweden

Dr. Z. Rigbi
Dept. of Chemical Engineering
Technion, Israel Institute
of Technology
Haifa 32000
Israel

Dr. Edmone Roffael
Wilhelm-Klauditz Institut für Holzforschung
33 Braunschweig
Bienroder Weg 54-E
West Germany

Dr. H. Sasaki
Composite Wood Section
Wood Research Institute
Kyoto University
Gokasho, Uji
Kyoto 611
Japan

Prof. H.P. Schreiber
Dept. of Chemical Engineering
Ecole Polytechnique
Campus de l'Université de Montréal
P.O. Box 6079, Station A
Montreal, Quebec H3C 3A7
Canada

Dr. J. Schultz
University de Haute-Alsace
C.N.R.S.
68200 Mulhouse
France

Dr. Harold Schonhorn
Bell Laboratories
Murray Hill
New Jersey 07974
U.S.A.

Dr. L.H. Sharpe
Bell Laboratories
Murray Hill
New Jersey 07974
U.S.A.

Dr. Kuo C. Shen
Forintek Canada Corporation
800 Montreal Road
Ottawa, Ontario K1G 3Z5
Canada

Dr. Dennis C. Smith
Faculty of Dentistry
University of Toronto
123 Edward Street
Toronto, Ontario M5G 1G6
Canada

Dr. P.R. Steiner
Forintek Canada Corporation
6620 West Marine Drive
Vancouver, B.C. V6T 1X2
Canada

Prof. Per Stenius
Director
The Swedish Institute for
Surface Chemistry
Drottning Kristinas VAG 45
S-11428 Stockholm
Sweden

Mr. John Talbott
Washington State University
Wood Technology Section
Pullman, Washington 99164
U.S.A.

Dr. J.D. Wellons
Dept. of Forest Products
Oregon State University
Corvallis, Oregon 97331
U.S.A.

Dr. G.E. Woodson
U.S.D.A.
Southern Forest Experimental Station
2500 Shreveport Highway
Pinesville, La. 71360
U.S.A.

Dr. Max Yan
Abitibi-Price Inc.
Sheridan Park
Mississauga, Ontario L5K 1A9
Canada

INDEX